T0140376

Conference Proceedings of the Society for Experimental Mechanics Series

Series Editor

Kristin B. Zimmerman
Society for Experimental Mechanics, Inc.,
Bethel, CT, USA

The Conference Proceedings of the Society for Experimental Mechanics Series presents early findings and case studies from a wide range of fundamental and applied work across the broad range of fields that comprise Experimental Mechanics. Series volumes follow the principle tracks or focus topics featured in each of the Society's two annual conferences: IMAC, A Conference and Exposition on Structural Dynamics, and the Society's Annual Conference & Exposition and will address critical areas of interest to researchers and design engineers working in all areas of Structural Dynamics, Solid Mechanics and Materials Research.

Ramin Madarshahian • Francois Hemez
Editors

Data Science in Engineering, Volume 9

Proceedings of the 40th IMAC, A Conference and Exposition on Structural Dynamics 2022

Editors
Ramin Madarshahian
Equifax
Boise, Idaho, USA

Francois Hemez
Lawrence Livermore National Security
Livermore, CA, USA

ISSN 2191-5644 ISSN 2191-5652 (electronic)
Conference Proceedings of the Society for Experimental Mechanics Series
ISBN 978-3-031-04124-2 ISBN 978-3-031-04122-8 (eBook)
https://doi.org/10.1007/978-3-031-04122-8

This Springer imprint is published by the registered company Springer Nature Switzerland AG
The registered company address is: Gewerbestrasse 11, 6330 Cham, Switzerland

Preface

Data Science in Engineering represents one of nine volumes of technical papers presented at the 40th IMAC, A Conference and Exposition on Structural Dynamics, organized by the Society for Experimental Mechanics, and held February 7–10, 2022. The full proceedings also include volumes titled *Nonlinear Structures & Systems*; *Dynamics of Civil Structures*; *Model Validation and Uncertainty Quantification*; *Dynamic Substructures*; *Special Topics in Structural Dynamics & Experimental Techniques*; *Rotating Machinery, Optical Methods & Scanning LDV Methods*; *Sensors and Instrumentation, Aircraft/Aerospace and Dynamic Environments Testing*; and *Topics in Modal Analysis & Parameter Identification*.

Data science in engineering is a new area of emphasis at the Society for Experimental Mechanics that focuses on the application of data analytics in structural and mechanical engineering. Machine learning, deep learning, big data, statistics, and related methods define the analytical toolset to process vast volumes of measurements and predictions, analyze complex phenomena, identify trends and relationships, and guide the development of predictive models through data. Advancements in sensing technologies (high-speed cameras, image processing, laser sensors, unmanned aerial vehicles, etc.) and high-performance computing increasingly require sophisticated frameworks to manage big data and extract useful information. Cloud systems are becoming unavoidable to store, classify, interpret, and visualize these data, raising security and privacy issues. Statistical and machine learning methods provide fast, resilient, adaptive, and scalable engines for the online monitoring of structures and mechanical systems, and to support decision-making and risk analysis.

The organizers would like to thank the authors, presenters, session organizers, and session chairs for their participation in this track.

Boise, Idaho, USA
Livermore, CA, USA

Ramin Madarshahian
Francois Hemez

Contents

Chapter 1
Model Updating for Nonlinear Dynamic Digital Twins Using Data-Based Inverse Mapping Models

Bas M. Kessels, Rob H. B. Fey, Mohammad H. Abbasi, and Nathan van de Wouw

Abstract In order to ensure that a digital twin accurately describes the dynamic behavior of its corresponding physical system, model updating is typically applied. This chapter introduces a (near) real-time method that uses inverse mapping models to update first-principles-based nonlinear dynamics models. The inverse mapping model infers a set of physically interpretable updating parameter values on the basis of a set of time-domain features extracted from measurements on the real system. Here, the inverse model is given by an artificial neural network that is trained using simulated data. By using a simple nonlinear multibody model, it is illustrated that this method is able to accurately and precisely update parameter values with low computational effort.

Keywords Model updating · Nonlinear dynamics · Digital twin · Machine learning

1.1 Introduction

As part of the fourth industrial revolution, digital twins have the potential to optimize system design, enable real-time monitoring and performance optimization, and facilitate predictive maintenance for a physical system. To achieve this potential, the digital twin has to be an accurate representation of its physical counterpart and remain so during its entire operational life. However, design uncertainties, manufacturing tolerances, material deterioration, and changing disturbance characteristics may cause the system dynamics to change. This inevitably leads to modeling errors and an initial first-principles model generally is not accurate enough to serve as a useful digital twin of the physical system in high-precision applications. Therefore, model updating techniques are employed to decrease the mismatch between the model and physical system. In this research, we focus on methodologies that update parameter values of first-principles models with fixed model structures, i.e., we have a reference model of which some updating parameter values are uncertain. Due to its application in an online, digital twin context, the model updating method should be: (1) computationally fast, (2) physically interpretable, and (3) applicable to complex nonlinear models. Often, traditional methods, e.g., sensitivity-based model updating and system identification techniques, however, do not meet these joint requirements, see, e.g., [4]. The method introduced in this chapter is based on the Neural Network Updating Method (NNUM) [2, 3], which uses an inverse mapping model to update, in near real time, model parameter values based on features of linear(ized) models. Although this methodology has been applied to linear systems in the literature, this chapter extends it to the case of nonlinear dynamical systems. Therefore, instead of traditionally used modal-domain features, we propose to use time-domain features.

B. M. Kessels (✉) · R. H. B. Fey · M. H. Abbasi
Department of Mechanical Engineering, Eindhoven University of Technology, Eindhoven, The Netherlands
e-mail: b.m.kessels@tue.nl; r.h.b.fey@tue.nl; m.h.abbasi@tue.nl

N. van de Wouw
Department of Mechanical Engineering, Eindhoven University of Technology, Eindhoven, The Netherlands

Department of Civil, Environmental, and Geo-Engineering, University of Minnesota, Minneapolis, MN, USA
e-mail: n.v.d.wouw@tue.nl

R. Madarshahian, F. Hemez (eds.), *Data Science in Engineering, Volume 9*, Conference Proceedings of the Society for Experimental Mechanics Series, https://doi.org/10.1007/978-3-031-04122-8_1

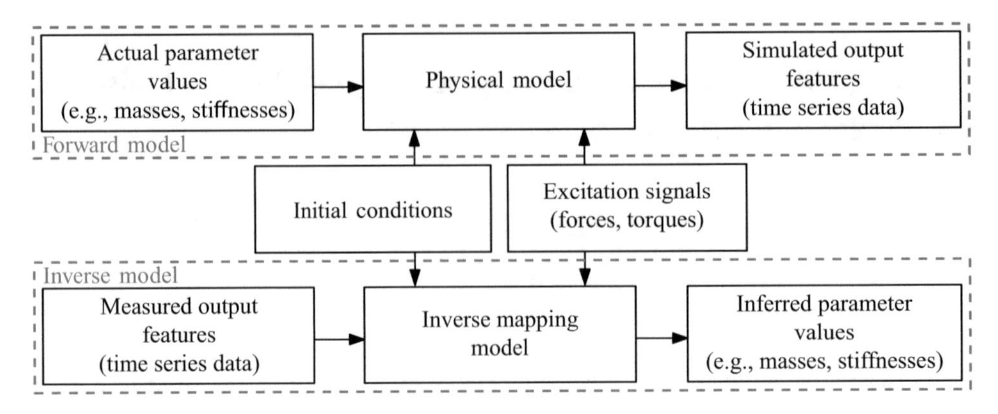

Fig. 1.1 Schematic representation of inputs and outputs to physical models and inverse models for model updating

1.2 Methodology

In the NNUM, physically interpretable parameters of a first-principles (forward) model are updated by using an inverse mapping model. In contrast to a forward dynamics model that, given some initial conditions and excitation signals, essentially maps a set of parameter values to a (simulated) output response, an inverse model does exactly the opposite. As shown in Fig. 1.1, the inverse model receives a set of (measured) output features, thereby capturing the behavior of the system, and maps these features to inferred parameter values that best approximate these features. In other words, parameterizing the reference model with the inferred parameter values and simulating this model yields simulated features similar to the measured features. Here, the reference model parameterized with the inferred parameter values thus serves as the updated model. As its name suggests, the NNUM uses an Artificial Neural Network (ANN) as the inverse model. To ensure that the ANN correctly maps the output features to parameter values, the ANN is trained using simulated training data that consists of numerous training samples of updating parameter values and their corresponding output features. These training data are generated offline by simulating the forward model for different combinations of parameter values and thereby extracting corresponding features from the simulated output signals. Once the ANN has been trained offline, in an online setting, measured output features are fed into the ANN yielding the inferred parameter values. Note that the initial conditions and excitation signals in both the offline and online phase in principle need to be identical, as these are implicitly learned by the ANN. Although the generation of training data requires a significant amount of computation time, a benefit of using ANN inverse models for updating is that, once an inverse model is available, (physically interpretable) parameters can be inferred in near real time. For linear models, extracted output features can consist of modal properties, e.g., eigenfrequencies, mode shapes, and antiresonance frequencies. Since a nonlinear system does not have these properties, this chapter proposes to extract features from time series data. Specifically, output samples at equidistant moments in time are used as output features.

1.3 Case Study: A Nonlinear Dynamic Multibody System

The above-introduced updating method is applied to a two-degrees-of-freedom nonlinear multibody system consisting of two connected rigid beams, shown schematically in Fig. 1.2. There are $n_p = 4$ updating parameters: masses m_1 and m_2 and spring constants k_y and k_θ, where it is assumed that their values may vary between the bounds of the admissible parameter space, $\mathbb{P} \subset \mathbb{R}^{n_p \times 1}$, specified in Table 1.1. The outputs of the system are the translation y of beam 1 and the rotation θ of beam 2. In this work, we want to demonstrate the proof of principle of the method, and, therefore, the "measured" output signals are simulated. Subsequently, these two simulated output signals are contaminated with artificial additive zero-mean Gaussian output noise with standard deviations $\sigma_y = 5 \times 10^{-5}$ m and $\sigma_\theta = 0.015$ rad, respectively.

For the output features, 10 equidistant time samples are used for each output, amounting to a total set of 20 features per (training) sample. To exemplify, the feature values corresponding to $\theta(t)$ are shown in Fig. 1.3. Furthermore, the initial conditions correspond to the system's static equilibrium position: $y_0 = 0.05$ m (undeformed spring length) and $\theta_0 \approx -0.21$ rad (due to gravity). The system is simultaneously excited by a force $F(t) = 0.15 \sin(0.3412t)$ N and torque $M(t) = 0.01 \cos(1.6635t)$ N·m. The frequencies of these excitations are chosen as the first and second eigenfrequency of the damped

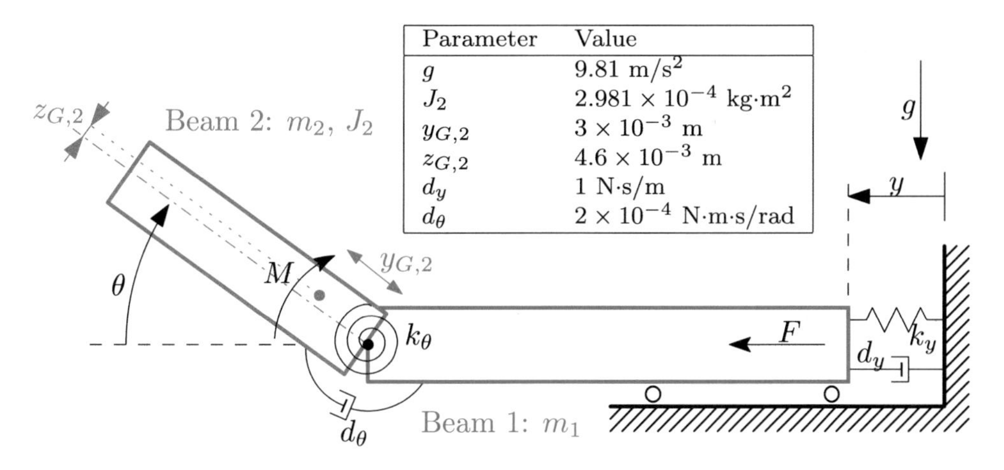

Fig. 1.2 Demonstrator model with corresponding non-updating parameters and their values

Table 1.1 Updating parameters and their corresponding lower and upper bound of the parameter space, bias, standard deviation, and mean absolute relative error

Parameter	Lower bound	Upper bound	μ_ϵ [%]	σ_ϵ [%]	$\mu_{\|\epsilon\|}$ [%]
m_1	1 kg	3 kg	−0.45	1.46	1.12
m_2	0.1 kg	0.2 kg	−0.17	2.53	1.99
k_y	5 N/m	15 N/m	−0.90	2.46	2.09
k_θ	0.027 N·m/rad	0.015 N·m/rad	0.28	0.71	0.61

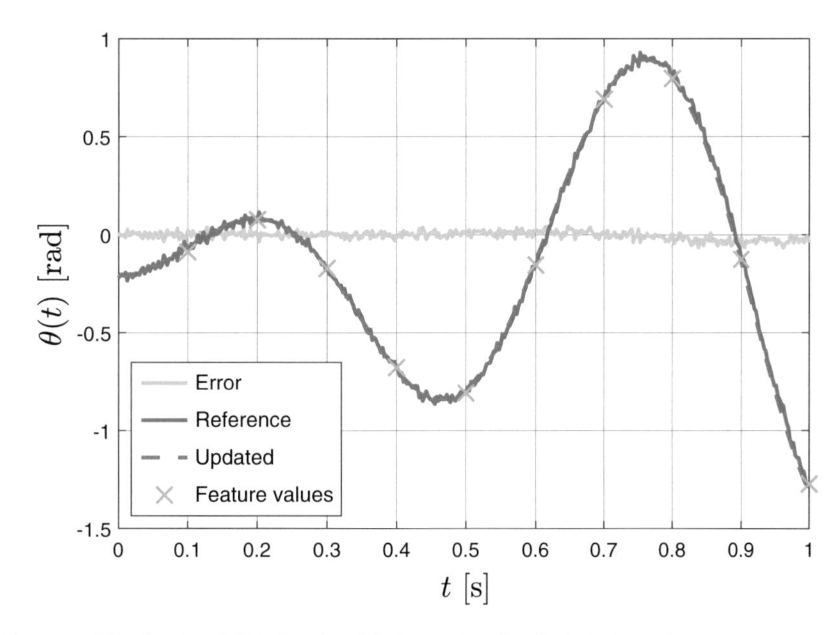

Fig. 1.3 Comparison of "measured" and updated $\theta(t)$ signal and feature values for single test sample

system linearized around the equilibrium position and parameterized with parameter values in the center of the admissible parameter space $\mathbb{P}$. This ensures that the output features are more sensitive to changes in the parameter values.

The ANN is trained using 10,000 training samples, generated for as many distinct combinations of updating parameter values, uniformly distributed in $\mathbb{P}$. Using a single core 2.60 GHz CPU, the offline data generation takes approximately 1500 s (elapsed time). To approximate reality, the training simulations are contaminated with noise with properties identical to the noise present in the "measured" signals. Furthermore, all parameter values and output features are normalized between 0 and 1. The utilized ANN is a fully connected direct feedthrough 5-layer (including input and output layer) perceptron network, where the subsequent layers have 20/ReLU, 40/ReLU, 40/ReLU, 20/ReLU, and 2/Sigmoid neurons/activation functions, where ReLU stands for Rectified Linear Unit. For more information about ANNs, the reader is referred to [1]. The ANN is trained in 150 epochs with a batchsize of 50 using the Adam optimizer, where the cost function is defined as the mean squared error. Training the ANN takes 47 s.

After training the ANN, performance of the updating strategy is evaluated using $n_t = 8000$ test samples (also uniformly distributed in $\mathbb{P}$). For each sample i, the set of output features is used to infer the set of updating parameter values, $\hat{\boldsymbol{p}}(i) \in \mathbb{P}$. Here, thanks to the computational simplicity of the trained ANN, parameter values of a single sample are inferred within approximately only 3 ms. The set of inferred parameter values is then compared to the set of parameter values, $\boldsymbol{p}(i)$, that was used to simulate the "measured" output features. This comparison is made by calculating the relative estimation error $\boldsymbol{\epsilon}_{\text{rel}}(i) \in \mathbb{R}^{n_p \times 1}$ for each sample:

$$\boldsymbol{\epsilon}_{\text{rel}}(i) = \left(\hat{\boldsymbol{p}}(i) - \boldsymbol{p}(i)\right) \oslash \boldsymbol{p}(i), \tag{1.1}$$

where $\oslash$ denotes the entrywise division operator. To assess the accuracy and precision of the inferred parameters, we use the following three metrics with respect to the relative error: the bias, $\boldsymbol{\mu}_\epsilon$, the standard deviation, $\boldsymbol{\sigma}_\epsilon$, and the mean absolute relative error, $\boldsymbol{\mu}_{|\epsilon|}$:

$$\boldsymbol{\mu}_\epsilon = \frac{1}{n_t}\sum_{i=1}^{n_t} \boldsymbol{\epsilon}_{\text{rel}}(i), \qquad \boldsymbol{\sigma}_\epsilon = \sqrt{\frac{1}{n_t - 1}\sum_{i=1}^{n_t} \left(\boldsymbol{\epsilon}_{\text{rel}}(i) - \boldsymbol{\mu}_\epsilon\right) \otimes \left(\boldsymbol{\epsilon}_{\text{rel}}(i) - \boldsymbol{\mu}_\epsilon\right)}, \qquad \boldsymbol{\mu}_{|\epsilon|} = \frac{1}{n_t}\sum_{i=1}^{n_t} \left|\boldsymbol{\epsilon}_{\text{rel}}(i)\right|, \tag{1.2}$$

where $\otimes$ denotes the entrywise multiplication operator. In Table 1.1, these error metrics are listed for each updating parameter separately. As indicated by these low error metrics, the updating parameters are inferred relatively accurately and precisely. Furthermore, simulating the updated model, i.e., the reference model parameterized with the inferred parameter values, results in an output signal almost identical to the measured signal; see Fig. 1.3.

1.4 Conclusions and Future Work

In this work, the NNUM is extended to nonlinear dynamical systems by exploiting time-domain output features. The resulting methodology, in which an inverse mapping model is used, is shown to enable near real-time updating of model parameters. For future work, we intend to investigate different types of time-domain features, use (optimal) excitation design to make the features more sensitive to parameter value changes, add quantification of the uncertainty in inferred parameter values, and apply this method to more complicated problems, e.g., simultaneously updating a larger number of parameters, and more complex models.

Acknowledgments This publication is part of the project Digital Twin (project 2.1) with project number P18-03 of the research program Perspectief, which is (mainly) financed by the Dutch Research Council (NWO).

References

1. Bishop, C.M.: Pattern Recognition and Machine Learning. Springer, Cambridge (2006). ISBN: 0-387-31073-8
2. Cooper, S.B., DiMaio, D.: Static load estimation using artificial neural network: application on a wing rib. Adv. Eng. Softw. **125**, 113–125 (2018)
3. Levin, R.I., Lieven, N.A.J.: Dynamic finite element model updating using neural networks. J. Sound Vib. **210**(5), 593–607 (1998)
4. Mottershead, J.E., Link, M., Friswell, M.I.: The sensitivity method in finite element model updating: a tutorial. Mech. Syst. Signal Process. **25**(7), 2275–2296 (2011)

Chapter 2
Deep Reinforcement Learning for Active Structure Stabilization

William Compton, Mason Curtin, Wilson Vogt, Alexander Scheinker, and Alan Williams

Abstract Structures are continuously exposed to external forcing, which can have dramatic variation both in loading magnitude and frequency content. A structure must implement control methods to protect both the structure and any contents from a wide variety of loadings, ranging from typical operational use to extreme cases such as natural disasters, including hurricanes or earthquakes. Current structures depend almost entirely on passive control systems; however, active controllers, which apply significant stabilizing forces to a structure, offer significant benefits, including improved performance, robustness to a wider range of input forcing, as well as the capability to be tuned or have their performance changed through software, instead of significant structural alterations. Traditional control systems, such as the PID controller, have key drawbacks: they require expert domain knowledge to tune, they can struggle to control high-order underactuated systems (which any high-fidelity structure model is guaranteed to be), and they rely on simple formulations of error or cost to minimize. Reinforcement learning provides a framework to learn high-performance control strategies directly from data, removing the necessity for expert domain knowledge. This work focuses on demonstrating the capabilities of data-driven methods in structural control by applying two deep reinforcement learning strategies (Soft Actor-Critic and Deep Deterministic Policy Gradient) to the control of a 3-story structure, modeled using a linear 3-degree-of-freedom mathematical model. The deep reinforcement learning controllers achieve 63% improvements over tuned PID controllers.

Keywords Reinforcement learning · Structure stabilization · Control

2.1 Introduction

Structure loading can take many forms, which can pose a threat to building integrity. Explicit design elements must be incorporated into structures to ensure the structure's response to any of its anticipated loadings – operation loadings, natural disasters, or environmental loadings – falls within safe bounds. Passive stabilizers are the main method for structural control used in modern structures, taking the form of either seismic isolation or energy dissipation devices [1]. The main benefits of passive designs include their independence from electricity, which can be called into question during a natural disaster, their relative simplicity when it comes to modeling and analyzing, and the wide base of historical implementations to draw on when implementing new designs. The inherent lack of ability to apply active forces to the structure limits passive controllers in both the types of loadings they can reject and their performance against any loading [2]. Improvements in performance to reject a wider range of disturbances can be made by adopting semi-active control devices, which modulate the response

W. Compton, M. Curtin and W. Vogt were equally contributing authors, alphabetical order is considered

W. Compton
Department of Mechanical Engineering, Georgia Institute of Technology, Atlanta, GA, USA

M. Curtin
Department of Mechanical and Aerospace Engineering, New Mexico State University, Las Cruces, NM, USA

W. Vogt
Department of Mechanical and Industrial Engineering, Montana State University, Bozeman, MT, USA

A. Scheinker · A. Williams (✉)
Applied Electrodynamics Group, Los Alamos National Laboratory, Los Alamos, NM, USA
e-mail: ascheink@lanl.gov; awilliams@lanl.gov

R. Madarshahian, F. Hemez (eds.), *Data Science in Engineering, Volume 9*, Conference Proceedings of the Society for Experimental Mechanics Series, https://doi.org/10.1007/978-3-031-04122-8_2

characteristics of a passive device based on sensor readings which measure the structure's response in real time. This allows the device to damp a wider variety of dynamic loadings [2]. Active structural control protocols have the significant advantage of being able to apply a counteractive force to the structure [3]. This allows for control of a wide range of input disturbances, as well as higher control performance. Additionally, active control systems can have their performance altered through software; changing the behavior of a passive or semi-active structural control system would require a significant retrofit of structural components.

This paper discusses the development of an active stabilizer implemented with reinforcement learning (RL) and trained in simulation on a validated 3-degree-of-freedom structure model.

2.2 Background

2.2.1 Reinforcement Learning

Reinforcement learning aims to maximize a numerical reward signal by mapping states to actions [4]. An RL agent learns about its environment by performing actions and observing their results, in terms of a reward function. The agents discussed in this paper are "model-free" agents, which learn purely from experience in the environment, without modeling it directly.

An RL problem is defined by: a state space, S, the set of all possible states of the agent; an action space, A, which is the set of all possible actions the agent can take in the environment; the agent's policy, $\pi(s) \to a$, which maps states to actions; and a reward function, $r(s, a)$, which computes the numerical reward the agent obtains for taking action a in state s.

When state or actions spaces are large or continuous, it quickly becomes infeasible to tabulate a policy, which maps actions to observed states. The use of neural networks to approximate a control policy has been shown to be very successful in many realms, including video games [5, 6]. Using neural networks as the basis for an RL agent is known as deep reinforcement learning (DRL). Based on prior work in DRL control [7–9], the algorithms Twin-Delayed Deep Deterministic Policy Gradient (TD3) and Soft Actor-Critic (SAC) are used in this setting. Both algorithms use a "Q-function" to approximate the maximum reward which can be obtained in the long term by selecting a specific action in the current state; this function is known as the Q-function and is mathematically defined:

$$Q\left(s_i, a_i\right) = r\left(s_i, a_i\right) + \sum_{t=t_i+1}^{t_\infty} \gamma^{t-t_0} r\left(s_t, \pi^*\left(s_t\right)\right) \tag{2.1}$$

where i denotes the sequence of temporal events; $\gamma \in (0, 1)$ is the discounting factor, decreasing the weight of future rewards; and π^* is the optimal action policy (generally unknown). The Bellman equation defines the Q-function recursively:

$$Q\left(s, a\right) = r\left(s, a\right) + \gamma Q\left(s', \pi^*\left(s'\right)\right) \tag{2.2}$$

By approximating this Bellman equation through experience, an RL agent can simply select the action which maximizes its Q-function for a particular state.

2.2.2 Data-Driven Control Methods Used on Dynamical Systems

Data-driven control strategies such as reinforcement learning have seen significant work in recent years. Eshkevari et al. addressed the problem of active structural control with reinforcement learning and were able to reduce the inter-story drift of a five-story building by 65% [10]. Jiang et al. controlled a linear discrete-time system with unknown dynamics used a reinforcement learning algorithm [11]. Silva and Neto used a Hamilton-Jacobi-Bellman approach to reconstruct system states without the use of a mathematical model, allowing for control of a nonlinear system without solving the nonlinear system at each timestep [12]. Each of these applications lacked any experimental work and generated data from mathematical models. Gao et al. used a physical system to compare the effectiveness of passive control, proportional-velocity control, and reinforcement learning control; their results revealed that reinforcement learning control was about 25% more effective in reducing the floor displacement and acceleration than passive control [13].

2.2.3 *Earthquake Vibrations*

Active control methods are only justified in the control of structures against earthquakes and other natural disasters; operational loadings can be dealt with effectively by passive means. This work will be concerned with controlling a structure during an earthquake, which generates a disturbance at the base of the structure. Typical ground motion excitations during an earthquake fall within the frequency range of 0.1–60 Hz, with intensity decreasing with distance from the epicenter [14]. This distribution is dependent upon earthquake size; large magnitude earthquakes typically excite in the frequency range 0.01–10 Hz, while smaller magnitudes are between 5 and 60 Hz [15]. Passive controllers struggle to maintain effectiveness over such a wide range of forcing frequencies, while active controllers can achieve strong performance over this whole range of vibrational excitation [1].

2.3 Methodology

This work will focus on controlling a 3-degree-of-freedom (3DoF) 3-story structure under base excitation, although the methodology is in no way specific to such a structure. This model was selected because authors have access to the physical structure for experimental investigations in future work. Structure responses are generated using a simulation of the structure derived from a 4DoF version of this model, originally developed by Nishio et al.; the reader is advised to refer to the original paper for details on the validation of the mathematical model [16]. To model an earthquake accurately, the 4DoF model presented by Nishio is converted to a 3DoF model by considering an input ground excitation described by x_1 and $\dot{x_1}$, rather than an input ground force, eliminating the degree of freedom related to the bottom layer. The building can be described by the mass, spring, and damper system shown in Fig. 2.1a.

The 3DoF system equations of motion are

$$[M]\{\ddot{x}\} + \left[C^{3DoF}\right]\{\dot{x}\} + [K]\{x\} = \{F(t, x_1, \dot{x_1})\} \tag{2.3}$$

where $[M]$ is the mass matrix, $[C^{3DoF}]$ is the damping matrix, and $[K]$ is the stiffness matrix. $\{F(t, x_1, \dot{x_1})\}$ is the force applied to the structure, and $\{x\}$ are the displacements at the nodal coordinates. The applied force is a function of the control force applied to the top floor, $f_4(t)$, and the trajectory of the bottom floor, x_1 and $\dot{x_1}$:

$$F(t, x_1, \dot{x_1}) = \begin{bmatrix} k_2 \\ 0 \\ 0 \end{bmatrix} x_1 - \begin{bmatrix} c_{21} \\ c_{31} \\ c_{41} \end{bmatrix} \dot{x_1} + \begin{bmatrix} 0 \\ 0 \\ 1 \end{bmatrix} f_4(t) \tag{2.4}$$

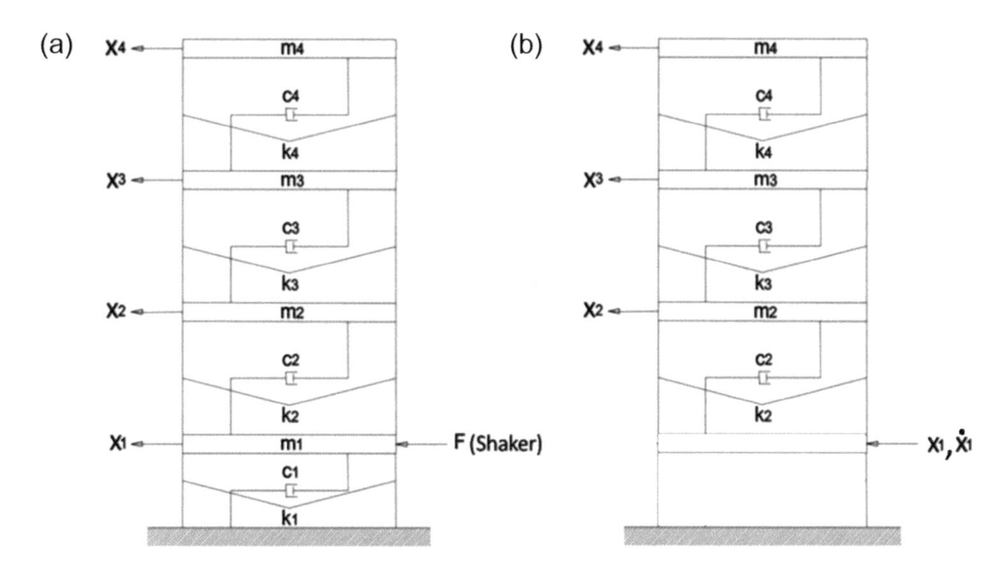

Fig. 2.1 Physical dynamic model of the original four DOF structure [15] (**a**) and adapted 3DoF structure (**b**)

The DRL agents are "model free"; they only select control forces and observe the data of the resulting response, without any explicit knowledge of the model of the system. Policy gradient and actor-critic methods are commonly used DRL techniques for learning in continuous state and action spaces such as this one [17]. Twin-Delayed Deep Deterministic Policy Gradient (TD3) and Soft Actor-Critic (SAC) are used in this work.

2.3.1 *Twin-Delayed Deep Deterministic Gradient (TD3)*

TD3 is a deterministic (maps states directly to actions), off-policy (learns from actions not taken in the current policy), model-free DRL algorithm, developed by Fujimoto et al. [18]. TD3 learns an "actor" network, π, which takes in an observation from the environment and outputs an action, and a "critic" network, Q, which outputs a Q-value estimate for a state-action pair. A replay buffer is used to "remember" past transitions between states, so that transitions can be sampled and learned from more than once, similarly to many other DRL implementations [5, 6, 19].

Pseudocode for TD3 is given in Appendix.

2.3.2 *Soft Actor-Critic (SAC)*

SAC is a stochastic (maps states to action probability distributions), off-policy, model-free DRL algorithm, proposed by Haarnoja et al. [20, 21]. SAC trains a random policy and uses a modified Q-function to simultaneously optimize expected reward and entropy, or randomness, of the learned policy. In order to maximize both long-term expected reward and policy entropy, the DRL problem reframes the Bellman update form of the Q-function as

$$Q\left(s,a\right)=r\left(s,a\right)+\gamma\left[Q\left(s',a'\right)-\alpha log\pi\left(a'|s'\right)\right] \quad (2.5)$$

where $\log\pi(a^{'}\,|\,s^{'})$ is the log probability of selection action $a^{'}$ from the policy distribution for state $s^{'}$. Double Q-learning is employed in SAC in the same fashion as TD3, with both Q-functions being regressed toward the target with mean squared error as the loss metric, using batches $\mathcal{B}$ sampled from the replay memory $\mathcal{R}$:

Pseudocode for SAC is given in Appendix.

2.3.3 *RL Agent Training and Excitation*

Reward functions $r(s,a)$ directly determine what elements of the structure's response the agent chooses to prioritize. Interstory drift (ISD), a displacement difference between two floors, and acceleration are two common rewards when analyzing structural control [22, 23]. Two reward functions are presented: R0 seeks to minimize top floor ISD and R1 minimizes a combination of all three ISDs and accelerations:

R0: $-\alpha_1|x_4-x_3|-\alpha_2 f_4$
R1: $-\alpha_1\left[|isd_4|+isd_4^2+|isd_3|+isd_3^2+|isd_2|+isd_2^2\right]-\alpha_2\left[|\ddot{x}_4|+\ddot{x}_4{}^2+|\ddot{x}_3|+\ddot{x}_3{}^2+|\ddot{x}_2|+\ddot{x}_2{}^2\right]-\alpha_3 f_4$

Two methods of structural excitation are considered: broadband excitation generates a random disturbance with frequency characteristics in the range [0.1, 100] Hz, and the impulse excitation delivers an initial impulse to the structure's base. Both excitations contain a wide range of frequency content and excite the building considerably.

DRL agents are compared to a classical PID controller to determine their relative effectiveness. The best PID gains were determined by a grid search, optimized for the particular disturbance type to yield a fair comparison.

2.4 Analysis

2.4.1 Comparative Analysis

The performance of the DRL agents and the PID controller are compared in their capability to control the 3DoF structure. Figure 2.2 shows the Reward 0 training curves of the TD3 and SAC algorithms. The reward is expressed as a percentage of the score achieved by the building with no control. The No-Control and PID lines are horizontal, because these responses do not change or improve with more data or episodes. In each case, the agents begin with heavy exploration and perform worse than the building without a controller; after fewer than 500 episodes, both DRL agents surpass the no-control score. By 1750 episodes, both agents obtain a higher score than the PID controller.

Table 2.1 shows the comparison evaluating the relative effectiveness of the different controllers. Fully trained agents, PID controller, and No-Control models are evaluated on a set of 10 disturbances; the evaluation metric used is average absolute ISD over the course of a trial. The results are averaged across the 10 trials. The PID results are reported as a percentage improvement over the No-Control score, while DRL agents are reported as percent improvements over both the No-Control and PID scores.

All three control methods are capable of dramatically reducing the interstory drift of structure under excitation. The DRL controllers were able to provide significant improvements over the PID controller as well, with TD3 beating the PID controller by 39% and SAC by 59% on average. SAC maintained a significant margin over TD3 algorithm. Figure 2.3 shows the ISD of the top floor over time to compare the structure response of the No-Control, PID, and SAC methods. Both the PID and SAC demonstrate significant reductions over the No-Control response; there are significant segments where the SAC controller holds the building extremely still, with drastic improvements over the PID controller.

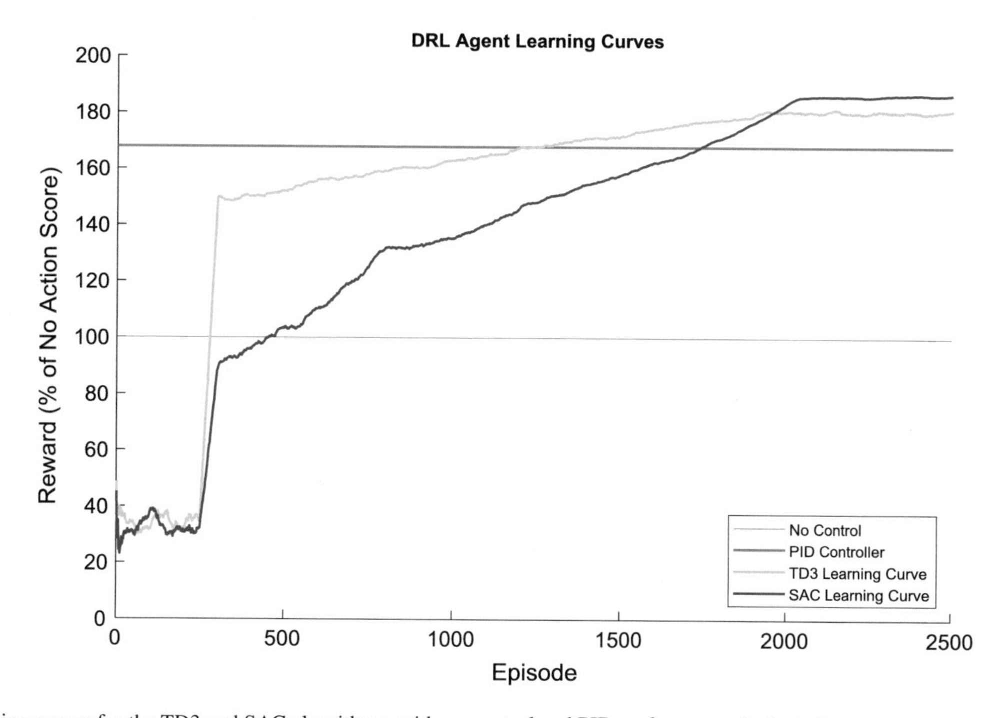

Fig. 2.2 Learning curves for the TD3 and SAC algorithms, with no-control and PID performances indicated

Table 2.1 PID vs. DRL Broadband Excitation Performance Comparison – Avg. Top Story ISD, Reward 0

Controller	% Imp. over No-Control	% Imp. over PID
PID	68%	–
TD3	81%	**39%**
SAC	87%	**59%**

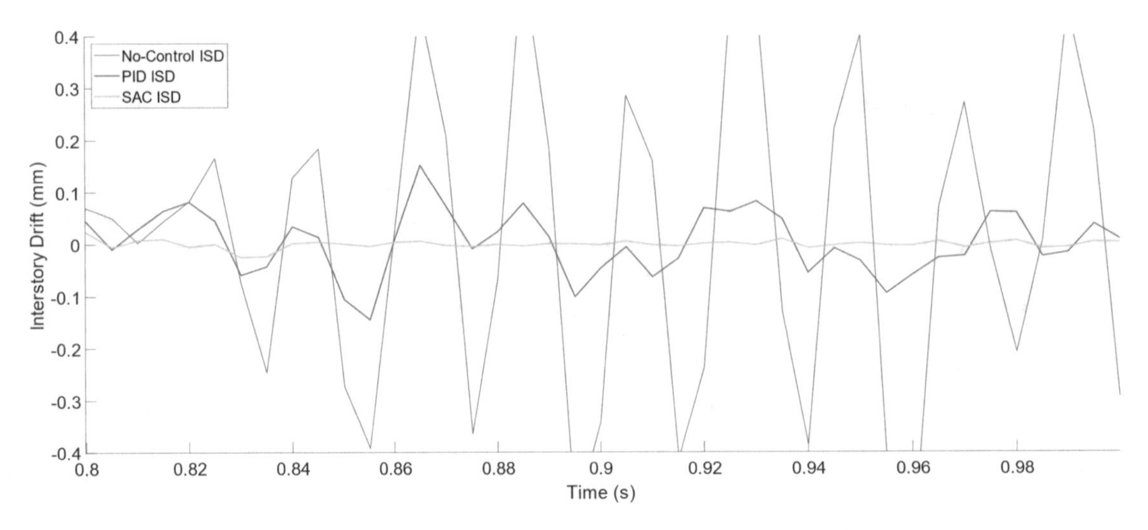

Fig. 2.3 Structure response comparison under various control strategies, subjected to broadband disturbance

Table 2.2 PID vs. DRL Impulse Excitation Performance Comparison – Avg. Top Story ISD, Reward 0

Controller	% Imp. over No-control	% Imp. over PID
PID[a]	72%	–
TD3	79%	**25%**
SAC	80%	**28%**

[a]The PID controller was re-tuned to specifically reject impulse disturbances

2.4.2 *Robustness to Variable Input*

For linear or locally linear systems, PID algorithms performance guarantees exist independently of the input disturbances given to the system. One main drawback of DRL controllers is significant difficulty associated with proving stability or performance, even for specific rather than general cases. Instead, to demonstrate robustness of the DRL controllers to variable input, the controllers previously trained on broadband excitation are evaluated on impulse excitation. This is a fundamentally different excitation and tests the ability of the DRL agent to generalize its control to unseen inputs. Table 2.2 summarizes the results of the DRL agents when deployed against impulse forcing.

Both TD3 and SAC agents were able to exceed 80% reductions in top story ISD over the No-Control response on average across the 10 tested trials. Additionally, the DRL agents were able to exceed the performance of the PID controller tuned to reject impulse disturbances by over 20% on average, indicating robustness to the type of input disturbance applied to the system. Figure 2.4 shows an example linear structure response to an impulse disturbance, in terms of top story ISD. The SAC controller was able to both achieve a lower peak ISD and drive ISD to zero quicker than the PID controller.

2.4.3 *Reward Function Analysis*

Altering the reward function used to train a DRL agent allows the agent's control objective to be easily customized. Reward 0, used in the above analysis, prioritizes only the top story ISD. Reward 1 instead prioritizes the ISD of all floors and acceleration of all floors. The DRL agents can customize their behavior to a reward function, unlike the PID controller, which can only minimize a single metric. Table 2.3 compares the PID and SAC controllers when SAC is trained with Reward 0 and Reward 1.

It is evident that the performance of Reward 0 comes at the cost of higher ISDs and accelerations of the other floors, as seen in Table 2.3. Changing to Reward 1, the SAC controller is still able to outperform the PID controller for top ISD, by 2%; however, it also outperforms the PID controller across all other metrics as well. Changing the reward function allows the DRL agents to prioritize different aspects of the building response, which provides a high degree of flexibility in their control capability.

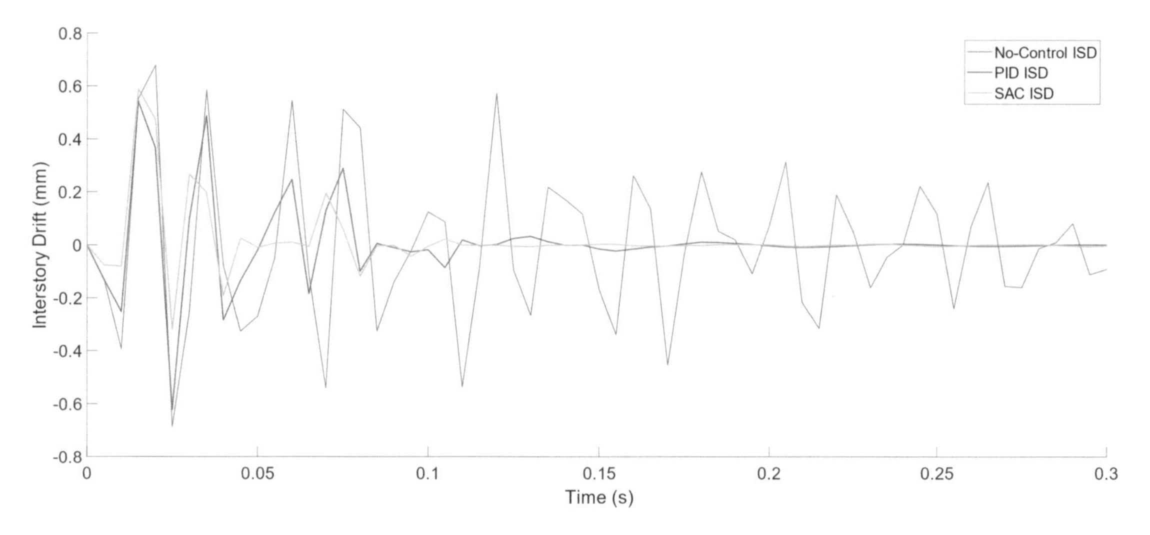

Fig. 2.4 Structure response comparison under various control strategies, subjected to impulse disturbance

Table 2.3 PID vs. SAC Broadband Excitation Performance Comparison – Reward 0, Reward 1

	Avg. Top ISD		Avg. All ISD		Avg. Top Accel.		Avg. All Accel.	
Controller	% Imp. over No-Control	% Imp. over PID	% Imp. over No-Control	% Imp. over PID	% Imp. over No-Control	% Imp. over PID	% Imp. over No-Control	% Imp. over PID
PID	68%	–	55%	–	23%	–	36%	–
SAC R0	87%	**39%**	46%	−18%	14%	−11%	21%	−23%
SAC R1	69%	**2%**	67%	**27%**	38%	**19%**	42%	**10%**

2.5 Conclusion

This paper presents a data-driven deep reinforcement learning (DRL) controller, capable of active structural control of a 3DoF building. The two algorithms investigated – Twin-Delayed Deep Deterministic Gradient (TD3) and Soft Actor-Critic (SAC) – both outperformed the No-Control and PID control responses on linear structures. When excited by a broadband disturbance, SAC decreased average ISD of the linear structure by 39% over the PID controller. It was also shown that by adjusting the reward functions of DRL controllers, different aspects of the building's response can be emphasized, such as floor acceleration. This is a level of adaptability PID controllers do not have. Also, the DRL agents can generalize their training to unseen disturbance types. The DRL agent achieved an 81% reduction in average top ISD over the No-Control when stabilizing impulse excitation, despite only being trained on broadband disturbances. Even when the PID controller was re-tuned to control impulse disturbances, the DRL agents consistently outperformed the PID controller in all evaluation categories. Overall, the DRL agents demonstrated strong performance for the purposes of active structural control.

Acknowledgments This research was funded by Los Alamos National Laboratory (LANL) through the Engineering Institute's Los Alamos Dynamics Summer School. The Engineering Institute is a research and education collaboration between LANL and the University of California San Diego's Jacobs School of Engineering. This collaboration seeks to promote multidisciplinary engineering research that develops and integrates advanced predictive modeling, novel sensing systems, and new development in information technology to address LANL mission replacement problems.

A.1 Appendix

Algorithm 1: Twin-Delayed Deep Deterministic Policy Gradient (TD3) Pseudocode

1. `Initialize: Policy` $\pi(s|\theta^{\pi})$, `Q-Functions` $Q_1\left(s, a|\theta_1^Q\right)$, $Q_2\left(s, a|\theta_2^Q\right)$, `Replay Memory` $\mathcal{R}$
2. `Set Target Parameters Equal to main parameters:`

$$\theta_{targ}^{\pi} \leftarrow \theta^{\pi}\ \theta_{1,targ}^{Q} \leftarrow \theta_{1}^{Q}\ \theta_{2,targ}^{Q} \leftarrow \theta_{2}^{Q}$$

```
3. REPEAT:
4.        Observe state s, select action
```

$$a(s) = clip\left(\pi\left(s|\theta^{\pi}\right) + \mathcal{N}\left(0, \sigma_1\right), a_{low}, a_{high}\right)$$

```
5.        Store Transition (s,a,r,s',d) in Replay Memory R
6.        IF s' is terminal: reset environment
7.        IF time to update THEN:
8.               Randomly sample batch B ~ R
9.               Compute Target Actions:
```

$$a'\left(s'\right) = \text{clip}\left(\pi_{targ}\left(s'|\theta_{targ}^{\pi}\right) + \mathcal{N}\left(0, \sigma_2\right), a_{min}, a_{max}\right)$$

```
10.              Compute Target Q-Values:
```

$$y = r\left(s, a\right) + \gamma\left(1 - d\right)\min_{i=1,2} Q_{i,targ}\left(s', a'\left(s'\right)\middle|\,\theta_{i,targ}^{Q}\right)$$

```
11.              Update Q-Functions:
```

$$L_{Q_i}\left(\theta_i^Q, \mathcal{B}\right) = \frac{1}{|\mathcal{B}|}\sum_{(s,a,r,s',d)\in\mathcal{B}}\left[Q_i\left(s, a|\,\theta_i^Q\right) - y\right]^2 i \in \{1, 2\}$$

```
12.              IF time to update policy, THEN:
13.                  Update Policy:
```

$$L_{\pi}\left(\theta^{\pi}, \mathcal{B}\right) = \frac{-1}{|\mathcal{B}|}\sum_{(s,a,r,s',d)\epsilon\mathcal{B}} Q_1\left(s, \pi\left(s|\,\theta^{\pi}\right)\middle|\theta_1^Q\right)$$

```
14.                  Update Target Networks:
```

$$\theta_{i,targ}^{Q} = \tau\theta_{i,targ}^{Q} + (1 - \tau)\,\theta_i^Q\ i \in \{1, 2\}$$

$$\theta_{targ}^{\pi} = \tau\theta_{targ}^{\pi} + (1 - \tau)\,\theta^{\pi}$$

Algorithm 2: Soft Actor-Critic (SAC) Pseudocode

```
1. Initialize: Policy π(s|θ^π), Q-Functions Q1(s,a|θ1^Q), Q2(s,a|θ2^Q), Replay Memory R
2. Set Target Parameters Equal to main parameters:
```

$$\theta_{1,targ}^{Q} \leftarrow \theta_{1}^{Q}\ \theta_{2,targ}^{Q} \leftarrow \theta_{2}^{Q}$$

```
3. REPEAT:
4.        Observe state s, select action
```

$$a(s) = \tanh\left(\mu + \sigma * \xi\right)\ \mu, \sigma = \pi\left(s|\theta^{\pi}\right), \xi \sim N\left(0, 1\right)$$

```
5.        Store Transition (s,a,r,s',d) in Replay Memory R
6.        IF s' is terminal: reset environment
7.        IF time to update THEN:
8.               Randomly sample batch B ~ R
```

9. `Compute Target Q-Values:`

$$y = r\left(s, a\right) + \gamma\left(1 - d\right)\left[\min_{i=1,2} Q_{i,targ}\left(s', \check{a}' \middle| \theta^{Q}_{i,targ}\right) - \alpha log\pi\left(\check{a}' | s', \theta^{\pi}\right)\right], \check{a}' \sim \pi\left(\bullet | s', \theta^{\pi}\right)$$

10. `Update Q-Functions:`

$$L_{Q_i}\left(\theta^{Q}_{i}, \mathcal{B}\right) = \frac{1}{|\mathcal{B}|}\sum_{(s,a,r,s',d)\in\mathcal{B}}\left[Q_i\left(s, a | \theta^{Q}_{i}\right) - y\right]^2 \; i \in \{1, 2\}$$

11. `IF time to update policy, THEN:`
12. `Update Policy:`

$$L_{\pi}\left(\theta^{\pi}, \mathcal{B}\right) = \frac{-1}{|\mathcal{B}|}\sum_{(s,a,r,s',d)\epsilon\mathcal{B}}\min_{i=1,2} Q_i\left(s, \check{a} | \theta^{Q}_{i}\right) - \alpha log\pi\left(\check{a} | s, \theta^{\pi}\right)$$

13. `Update Target Networks:`

$$\theta^{Q}_{i,targ} = \tau\theta^{Q}_{i,targ} + \left(1 - \tau\right)\theta^{Q}_{i} \; i \in \{1, 2\}$$

References

1. Saaed, T.E., Nikolakopoulos, G., Jonasson, J.-E., Hedlund, H.: A state-of-the-art review of structural control systems. J. Vib. Control. **21**(5), 919–937 (2015). https://doi.org/10.1177/1077546313478294
2. Cheng, F.Y.: Smart Structures: Innovative Systems for Seismic Response Control. CRC Press (2008)
3. Yao, J.T.: Concept of structural control. J. Struct. Div. **98**(7), 1567–1574 (1972)
4. ROBOT LEARNING, edited by Jonathan H. Connell and Sridhar Mahadevan, Kluwer, Boston, 1993/1997, xii+240 pp., ISBN 0-7923-9365-1 (Hardback, 218.00 Guilders, $120.00, £89.95). (1999). Robotica **17**(2), 229–235. https://doi.org/10.1017/S0263574799271172
5. Mnih, V., Kavukcuoglu, K., Silver, D., Graves, A., Antonoglou, I., Wierstra, D., Riedmiller, M.: Playing atari with deep reinforcement learning. arXiv preprint arXiv:1312.5602 (2013)
6. Mnih, V., Kavukcuoglu, K., Silver, D., Rusu, A.A., Veness, J., Bellemare, M.G., et al.: Human-level control through deep reinforcement learning. Nature. **518**(7540), 529–533 (2015)
7. Chu, Z., Sun, B., Zhu, D., Zhang, M., Luo, C.: Motion control of unmanned underwater vehicles via deep imitation reinforcement learning algorithm. IET Intell. Transp. Syst. **14**(7), 764–774 (2020)
8. Dankwa, S., Zheng, W.: Twin-delayed DDPG: a deep reinforcement learning technique to model a continuous movement of an intelligent robot agent. In: Proceedings of the 3rd international conference on vision, image and signal processing, pp. 1–5 (2019, August)
9. Haarnoja, T., Zhou, A., Hartikainen, K., Tucker, G., Ha, S., Tan, J., ... Levine, S.: Soft actor-critic algorithms and applications. arXiv preprint arXiv:1812.05905 (2018)
10. Eshkevari, S.S., Eshkevari, S.S., Sen, D., Pakzad, S.N. RL-Controller: a reinforcement learning framework for active structural control. arXiv preprint arXiv:2103.07616 (2021)
11. Jiang, Y., Kiumarsi, B., Fan, J., Chai, T., Li, J., Lewis, F.L.: Optimal output regulation of linear discrete-time systems with unknown dynamics using reinforcement learning. IEEE Trans. Cybern. **50**(7), 3147–3156 (2019)
12. Da Silva, F.N., Neto, J.V.D.F.: Data driven state reconstruction of dynamical system based on approximate dynamic programming and reinforcement learning. IEEE Access. **9**, 73299–73306 (2021)
13. Gao, H., He, W., Zhang, Y., Sun, C.: Vibration control based on reinforcement learning for a flexible building-like structure system with active mass damper against disturbance effects. In: 2020 59th IEEE Conference on Decision and Control (CDC), pp. 2380–2385. IEEE (2020, December)
14. Hays, W.W. (ed.): Facing Geologic and Hydrologic Hazards: Earth-Science Considerations, vol. 1240. US Department of the Interior, Geological Survey (1981)
15. Tosi, P., Sbarra, P., De Rubeis, V.: Earthquake sound perception. Geophys. Res. Lett. **39**(24) (2012)
16. Nishio, M., Farrar, C., Hemez, F., Stull, C., Park, G., Cornwell, P., et al.: Feature Extraction for Structural Dynamics Model Validation (No. LA-UR-16-20151). Los Alamos National Lab. (LANL), Los Alamos (2016)
17. Lillicrap, T.P., Hunt, J.J., Pritzel, A., Heess, N., Erez, T., Tassa, Y., ... Wierstra, D.: Continuous control with deep reinforcement learning. arXiv preprint arXiv:1509.02971 (2015)
18. Fujimoto, S., Hoof, H., Meger, D.: Addressing function approximation error in actor-critic methods. In: International conference on machine learning, pp. 1587–1596. PMLR (2018, July)
19. Lin, L.J.: Reinforcement Learning for Robots Using Neural Networks. Carnegie Mellon University (1992)
20. Haarnoja, T., Zhou, A., Abbeel, P., Levine, S.: Soft actor-critic: off-policy maximum entropy deep reinforcement learning with a stochastic actor. In: International Conference on Machine Learning (pp. 1861–1870). PMLR (2018, July)

21. Haarnoja, T., Zhou, A., Hartikainen, K., Tucker, G., Ha, S., Tan, J., ..., Levine, S.: Soft actor-critic algorithms and applications. arXiv preprint arXiv:1812.05905 (2018)
22. Casciati, S., Chen, Z.: An active mass damper system for structural control using real-time wireless sensors. Struct. Control Health Monit. **19**(8), 758–767 (2012)
23. Skolnik, D.A., Wallace, J.W.: Critical assessment of interstory drift measurements. J. Struct. Eng. **136**(12), 1574–1584 (2010)

Chapter 3
Estimation of Structural Vibration Modal Properties Using a Spike-Based Computing Paradigm

Jabari Allen, Raymond Chu, Troy Sims, Alessandro Cattaneo, Gregory Taylor, Andrew Sornborger, and David Mascareñas

Abstract Spiking neural networks are an emerging concept that draws inspiration from the computational neuroscience research community. Spiking neural networks combine spike-based computing and machine-learning-based neural networks that emulate the operation of the human brain. Spiking neural networks have the ability to be easily integrated into neuromorphic hardware, such as Intel's *Loihi* chip. The advantages of neuromorphic hardware are its high-speed computation and low-power consumption in comparison to traditional electronics. These factors have an important role for the future of smart systems and the reliability of structural health monitoring. Currently, coupling spike-based computing to continuous-valued signals, which are typically measured in structural dynamics, is rare. This paper aims to explore spiking neural networks and their possible application in structural dynamics and modal analysis using Nengo, a large-scale neural network simulation package. In this work, we implement output-only modal identification techniques that rely on solving the blind source separation problem using spike neural networks to extract the natural frequencies, mode shapes, and damping ratios of a simulated structural system being exposed to dynamic loading.

Keywords Neuromorphic processing · Modal analysis · Dynamics

3.1 Introduction

Contact vibration measurement sensors require costly installation and maintenance while only capturing discrete measurements. Traditional frame-based cameras are a step up from traditional sensors in that cameras are more affordable and agile and produce high spatial resolution [1–5]. However, these cameras also report redundant information because they capture entire frames of data rather than the dynamic changes. Silicon retinas, also called dynamic vision sensors (DVS) or neuromorphic cameras, are the most recent event-based technology that provides all of the benefits of traditional cameras while omitting redundant data. This is done by imitating the biological vision organ, where each neuron functions and fires signals asynchronously from another. The result is that only changes in local intensity or brightness are reported. The current challenges of neuromorphic cameras are that new algorithms must be developed in order to process the data obtained from event cameras.

J. Allen
Department of Electrical and Computer Engineering, FAMU-FSU College of Engineering, Florida A&M University, Tallahassee, FL, USA

R. Chu
Department of Structural Engineering, Jacobs School of Engineering, University of California San Diego, La Jolla, CA, USA
e-mail: rzchu@ucsd.edu

T. Sims
Department of Electrical Engineering, New Mexico Institute of Mining & Technology, Socorro, NM, USA
e-mail: troy.sims@student.nmt.edu

A. Cattaneo (✉) · A. Sornborger · D. Mascareñas
Los Alamos National Laboratory, Los Alamos, NM, USA
e-mail: cattaneo@lanl.gov; sornborg@lanl.gov; dmascarenas@lanl.gov

G. Taylor
Department of Mechanical and Aerospace Engineering, New Mexico State University, Las Cruces, NM, USA
e-mail: gtaylor3@nmsu.edu

R. Madarshahian, F. Hemez (eds.), *Data Science in Engineering, Volume 9*, Conference Proceedings of the Society for Experimental Mechanics Series, https://doi.org/10.1007/978-3-031-04122-8_3

Spike-based computing is a form of artificial neural networks (ANN) that offer low power consumption and high computational speed. Applications of ANNs include but are not limited to structural system identification, damage identification, performance evaluation, and structural design optimization [6]. However, due to its novel emergence into the realm of neuromorphic processing, there is little-to-no work on coupling spike-based computational pipelines to continuous-valued signals that are typically measured in structural dynamics. This study will fill the gap in the context of structural vibration mode shapes.

3.1.1 Blind Source Separation

Blind source separation (BSS) is often referred to as the cocktail party problem, where several people are speaking and information wants to be extracted from only one person. This same problem appears in several fields where multivariate data appears, such as financial time series and astrophysical datasets. It also appears in structural dynamics where damage detection and health monitoring are needed [7]. BSS techniques include independent component analysis (ICA), second-order blind identification (SOBI), and the complexity pursuit algorithm (CP). These techniques are used to be able to extract modal shapes and different parameters from small and large-scale structures when doing output-only modal analysis [1, 7–10].

Overall, the goal of BSS is to recover unobserved source signals from their observed mixtures. To do this, general assumptions have to be made about the source signals. The challenges that appear in structural dynamics are that the mechanical systems response is time-based and relates to a convolution mixture excitation instead of a simpler static mixture [7]. In addition, the mixing matrix of different BSS methods is only "identifiable from the scaling and permutation of the rows." The result of this means that the order and variance of the source signals cannot be known [7, 8].

ICA is one of the most popular BSS techniques. ICA assumes that the observed data is a linear combination of independent sources. It also should be noted that ICA is superior to PCA when the focus is on feature extraction [7]. A more in-depth tutorial is found that explains the mathematics and certain challenges and workarounds when using ICA [11]. ICA does have notable drawbacks where it is restricted to only undamped and very lightly damped structures. However, it is still an efficient method except on high temporal structures [7, 8].

SOBI tries to take advantage of the temporal structure of sources to facilitate their separation; the math is further explained in the source paper. In addition, SOBI is applied to estimate the mixing matrix and the sources; the natural frequencies and damping ratios are then extracted for the different vibration modes. Overall, SOBI works robustly when damping effects and forced responses are used on a structure [7].

The CP algorithm is computationally efficient and user-friendly. It also can be applied on modes that are closely spaced and structures that are highly damped. The CP algorithm is further explained mathematically. The CP method was able to extract the modal shapes from an experimental model, which proves its effectiveness [8]. In addition, the CP method was applied on image data. The image data underwent motion magnification to accentuate the motion in the video. Blind identification was needed because the structures don't have a uniform mass distribution, and, as result of this, the principal components still consisted of a mixture of the modes. The CP method is validated by laboratory experiments on a bench-scale building structure and a cantilever beam. The results of the experiments showcase that the CP method works well [9, 10].

Overall, these sources help in understanding the different BSS techniques that are used, as well as the background, challenges, and benefits of each technique that is introduced. The sources also help with motivation for output-only modal analysis and insight into the data that will need to be represented in Nengo [1, 7–10].

3.1.2 Neural Hardware

Conventional computers are organized around a centralized processing architecture, known as the von Neumann model, which is suited to run sequential, digital, procedure-based programs. A fundamental limitation of the von Neumann model is the clear bounds of a processing unit physically separated from a storage unit that is connected through a bus for data transfer. Frequent movement of data between a faster processing unit and a slower memory unit through this bandwidth-constrained bus leads to a "memory wall bottleneck" that limits computing throughput and energy efficiency. Such an architecture is inefficient for computational models that are distributed, massively parallel, and adaptive, like those commonly used for neural networks in artificial intelligence [12, 13].

Modeled after the brain, neuromorphic hardware systems that implement neuronal and synaptic computations through spike-driven communication may enable energy-efficient machine intelligence. In the context of information processing, the former generation of deep learning networks (DLNs) use real-valued computation, like the amplitude of a signal, whereas a spiking neural network (SNN) uses the timing of the signals called spikes that are essentially binary events, either 0 or 1, to process information. In a spike-based approach, SNNs are trained using timing information and offer the obvious advantages of sparsity and efficiency in overall spiking dynamics [12].

3.2 Methods

Blind source separation problems have several different traditional approaches to solving them: projection pursuit (PP), independent component analysis (ICA), principal component analysis (PCA), and complexity pursuit (CP). By decomposing each of these approaches to their subsystems, these subsystems can be rebuilt into a spike-based computational paradigm using Nengo and tested to ensure their proper functionality. These spike-based subsystems can then be connected to reconstruct their respective BSS approach into a spike-based computing paradigm.

3.2.1 *Gradient Ascent*

Gradient ascent is a search strategy used to maximize different measures for blind source separation techniques, for example, non-Gaussianity for PP and independence for ICA [11]. The implementation of gradient ascent in PP is finding the rate of change of the local kurtosis with respect to each unmixing coefficient. The direction of steepest ascent is desired because PP seeks to maximize kurtosis. The implementation of gradient ascent into Nengo will allow for a spike-based computational network for PP.

3.2.2 *Independent Component Analysis/Principal Component Analysis*

Both ICA and PCA are statistical transformations and are used for blind source separation and feature extraction. PCA finds the principal components of a dataset. Principal components correspond to directions with greatest variance in data with each succeeding step finding the direction that explains most variance. ICA assumes there exists independent signals and seeks to address higher order dependence with the goal of recovering the original signals from a signal mixture. Generally, data is preprocessed before applying ICA to remove correlation called “whitening” with PCA being one way to whiten signals [11, 14].

3.2.3 *Oja’s Learning Rule*

Oja’s learning rule, or Oja’s rule, is a mathematical formalization of the Hebbian learning rule and is a model of how neurons in the brain or in artificial neural networks change connection strength, or learn, over time. A mathematical analysis of the Oja’s learning rule shows that if the weights converge in the Oja’s learning rule, then the weight vector becomes one of the eigenvectors of the input covariance matrix, and the output of the neuron becomes the corresponding principal component. Principal components are defined as the inner products between the eigenvectors and the input vectors, and, for this reason, the simple neuron learning by the Oja’s rule becomes a principal component analyzer. Similarly, small changes in the Oja’s rule can produce independent, instead of principal, components in such a case and have been shown to give one of the independent hidden factors under suitable assumptions to perform ICA [15].

3.2.4 Data Acquisition

To properly test the different spike-based BSS methods, a simulation of a multiple-degree-of-freedom (MDOF) structure with base excitation and damping, seen in Fig. 3.1, was created by solving the structural dynamic equations of motion seen in Eq. 3.1 [16].

$$\ddot{x} + C\dot{x} + Kx = -M\Gamma\ddot{x}_g \tag{3.1}$$

The structural simulation was created in MATLAB. These equations were solved using the Newmark-beta numerical integration method. The equations that were used can be seen in Eqs. 3.2, 3.3, 3.4, 3.5 and 3.6 [17].

$$\hat{K} = K + \left(\frac{1}{\beta\left(\Delta t^2\right)}\right)M + \left(\frac{\alpha}{\beta\left(\Delta t\right)}\right)C \tag{3.2}$$

$$\hat{F} = F_{i+1} + \left(\left(\frac{1}{\beta\left(\Delta t^2\right)}\right)M + \left(\frac{\alpha}{\beta\left(\Delta t\right)}\right)C\right)x_i + \left(\left(\frac{1}{\beta\left(\Delta t\right)}\right)M + \left(\frac{\alpha}{\beta} - 1\right)C\right)\dot{x}_i + \left(\left(\frac{1}{2\beta} - 1\right)M + (\Delta t)\left(\frac{\alpha}{2\beta} - 1\right)C\right)\ddot{x}_i \tag{3.3}$$

$$x_{i+1} = \hat{K}^{-1}\hat{F} \tag{3.4}$$

$$\dot{x}_{i+1} = \left(\frac{1}{\beta\left(\Delta t\right)}\right)(x_{i+1} - x_i) + \left(1 - \frac{\alpha}{\beta}\right)\dot{x}_i + (\Delta t)\left(1 - \frac{\alpha}{2\beta}\right)\ddot{x}_i \tag{3.5}$$

$$\ddot{x}_{i+1} = \left(\frac{1}{\beta\left(\Delta t^2\right)}\right)(x_{i+1} - x_i) + \left(\frac{1}{\beta\left(\Delta t\right)}\right)\dot{x}_i + \left(\frac{1}{2\beta} - 1\right)\ddot{x}_i \tag{3.6}$$

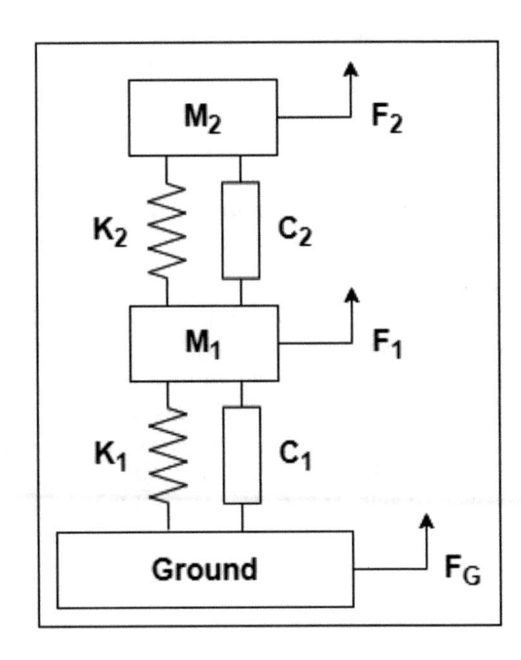

Fig. 3.1 MDOF system with damping and base excitation

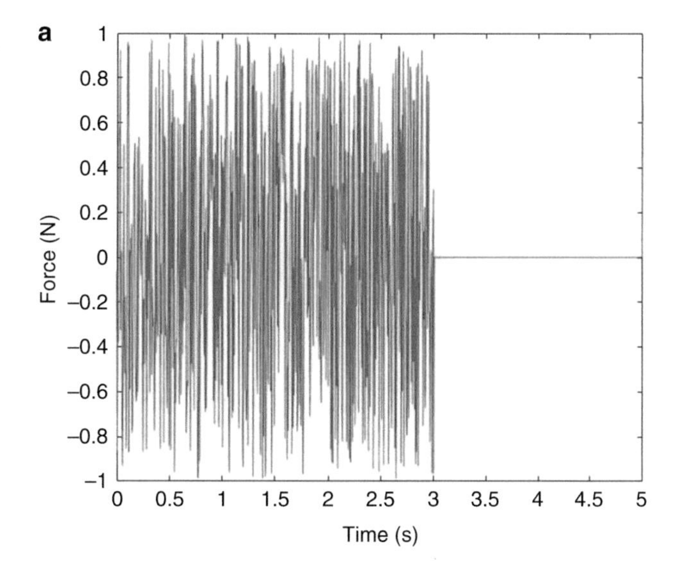

Fig 3.2a Force applied on ground

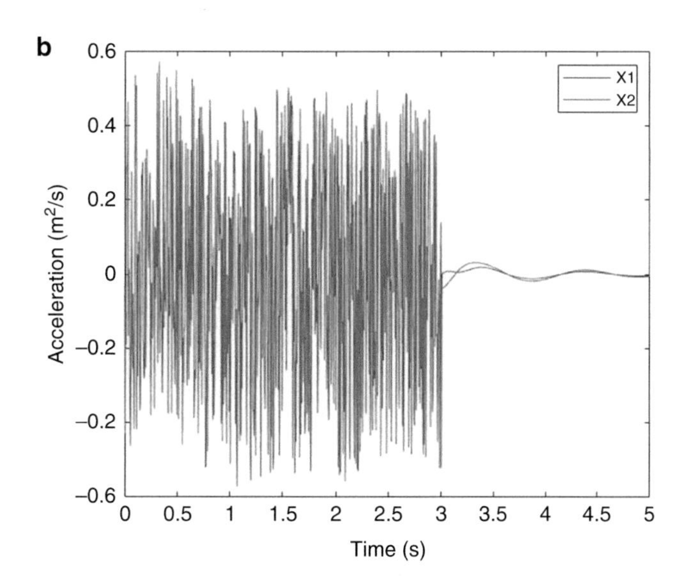

Fig 3.2b Acceleration vs. time 2DOF system

This structural simulation is useful because the modal properties are known and allows for an easy comparison between the actual and expected values of the modal properties of the structure. This structural simulation takes the force applied to ground as input (Fig. 3.2a) and outputs acceleration (Fig. 3.2b), velocity (Fig. 3.2c) and position (Fig. 3.2d) at each mass location. The acceleration data of the simulation can be used to replicate the data that would be acquired from a physical experiment and will be fed into the different spike-based BSS methods [18–20].

3.3 Analysis

3.3.1 Revised Sanger's Learning Rule

An issue that was found with Sanger's learning rule, which is an extension of the Oja's learning rule, is that the large singular value diverges for a high learning rate, and the low singular value converges slowly for low learning rates, creating the need for the learning rate to be adaptive. To fix this, the learning rate should be proportional to the singular value, but this will

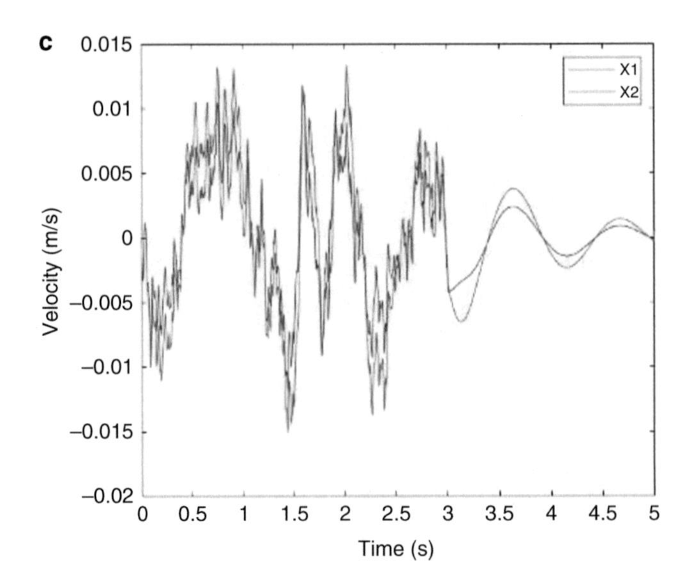

Fig 3.2c Velocity vs. time, 2DOF system

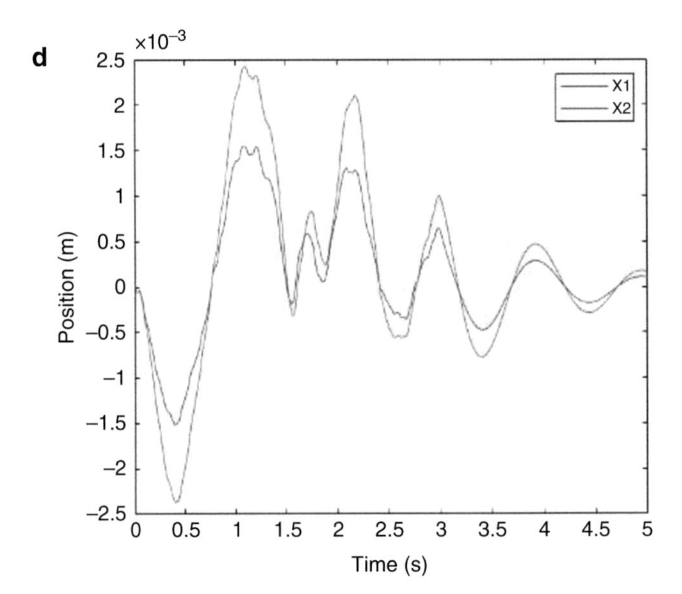

Fig 3.2d Position vs. time, 2DOF system

only work when the vector is aligned with the singular values. To overcome this, the L1 norm of the weighted correlation matrix is used, and we get the following revised learning rule:

$$\eta = \frac{\eta_0}{\left| W(n)^T X X^T W(n) \right|_1}$$

In Fig. 3.3a, it is shown that when the learning rate is not adaptive, the second and third principal components are not able to converge completely. However, when the learning rule uses an adaptive learning rate in Fig. 3.3b, the second and third principal components converge completely.

The revised Sanger's learning rule can be successfully applied on the MDOF undamped simulation structure that had uniform masses and spring constants. As a result, the eigenvectors of the simulation are obtained and then normalized. The mode shapes that were extracted from the revised Sanger's learning rule were then compared to the eigenvectors of the known mass and stiffness matrix. Figure 3.4 showcases the comparison of the mode shapes from the actual and estimated mode shapes.

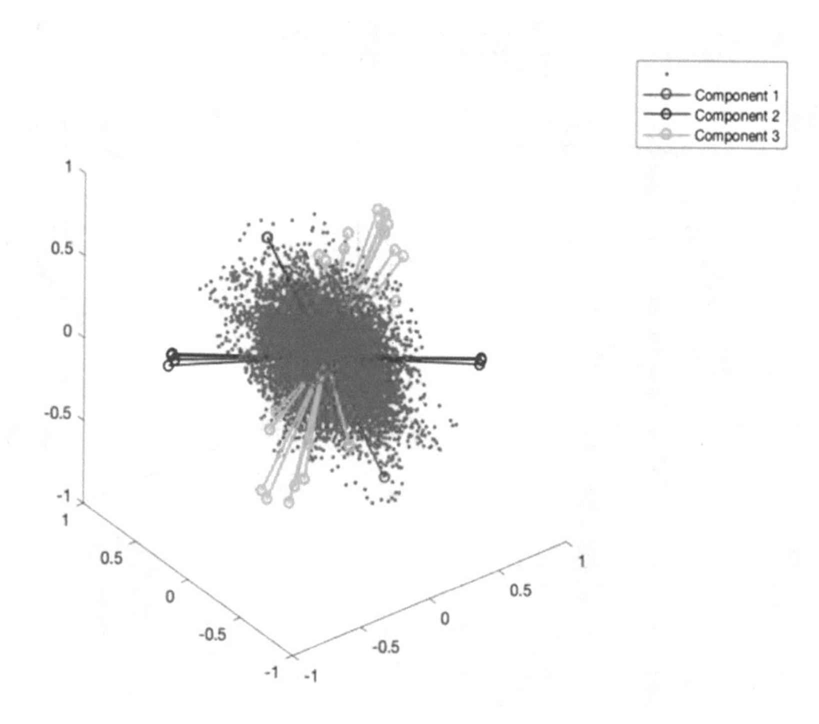

Fig 3.3a Slow convergence of Sanger's rule

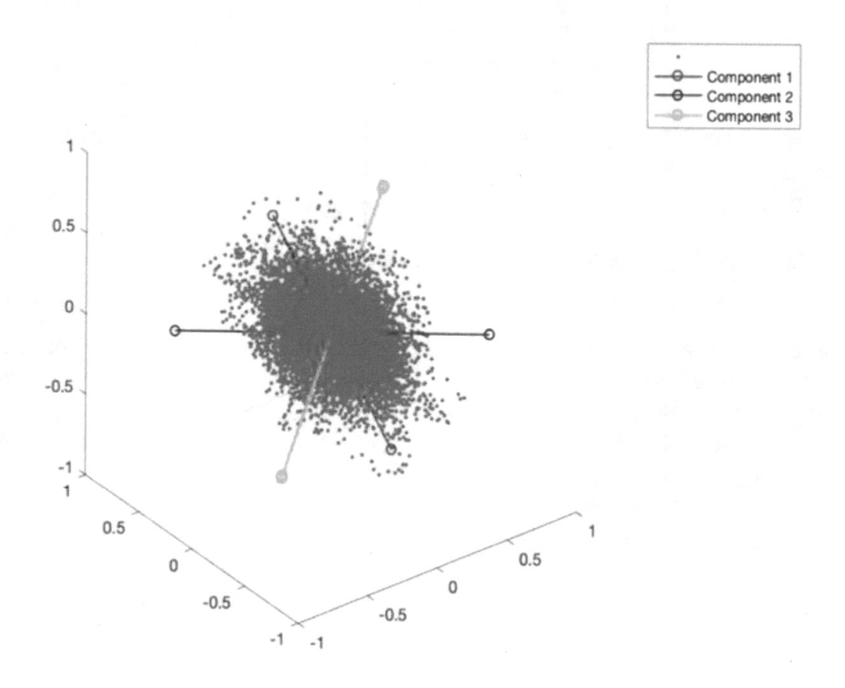

Fig 3.3b Convergence of revised Sanger's rule

Next, the revised Sanger's learning rule was applied on a physical experiment dataset, which also consisted of uniform masses and spring constants (see Fig. 3.5). Similarly, eigenvectors of the physical system were also obtained. These extracted mode shapes were compared to the eigenvectors of the analytical model, which only is possible due to both datasets coming from a MDOF structure that had uniform masses and springs. Additionally, PCA and ICA were applied using conventional digital computing techniques. This was applied on both the simulation and physical datasets.

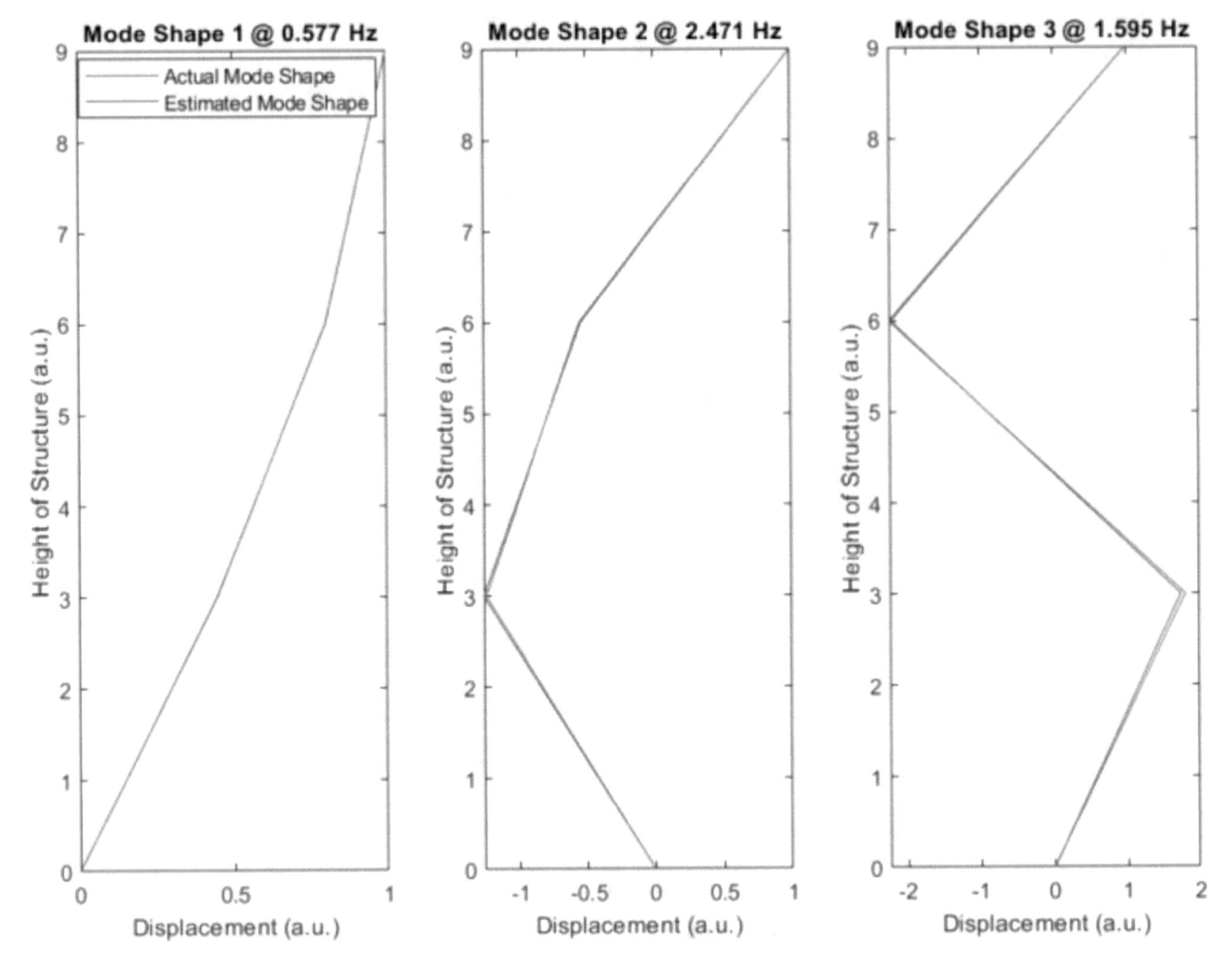

Fig. 3.4 Mode shapes of MDOF undamped simulation using revised Sanger's rule

To further verify experimental results, the modal assurance criterion (MAC) was applied on the different techniques: analytical, conventional, and revised Sanger's learning rule. The conventional and revised Sanger's learning rule were self-analyzed with the MAC matrix to ensure that the results are expected, shown in Figs. 3.6a and 3.6b. Furthermore, the conventional technique was compared to the analytical technique, shown in Fig. 3.6c. Lastly, the conventional technique was compared to the revised Sanger's learning rule using the MAC, shown in Fig. 3.6d. Overall, the mode shapes align with the expected mode shapes in the MAC, showcasing that the revised Sanger's learning rule can accurately extract the mode shapes of the physical and simulation structure datasets.

3.4 Conclusion

Our scientific contribution has been the application of PCA using the revised Sanger's learning rule. This allowed for the extraction of mode shapes of both physical and simulation structures, consisting of uniform masses and spring constants, using a neural network approach. Future work for this project would include the implementation of the revised Sanger's learning rule into a large-scale spiking neural network platform, the ability to extract natural frequencies of a structure using the revised Sanger's learning rule, implementing the other BSS methods (ICA and complexity pursuit), and implementing these computational methods on neuromorphic devices, such as silicon retinas.

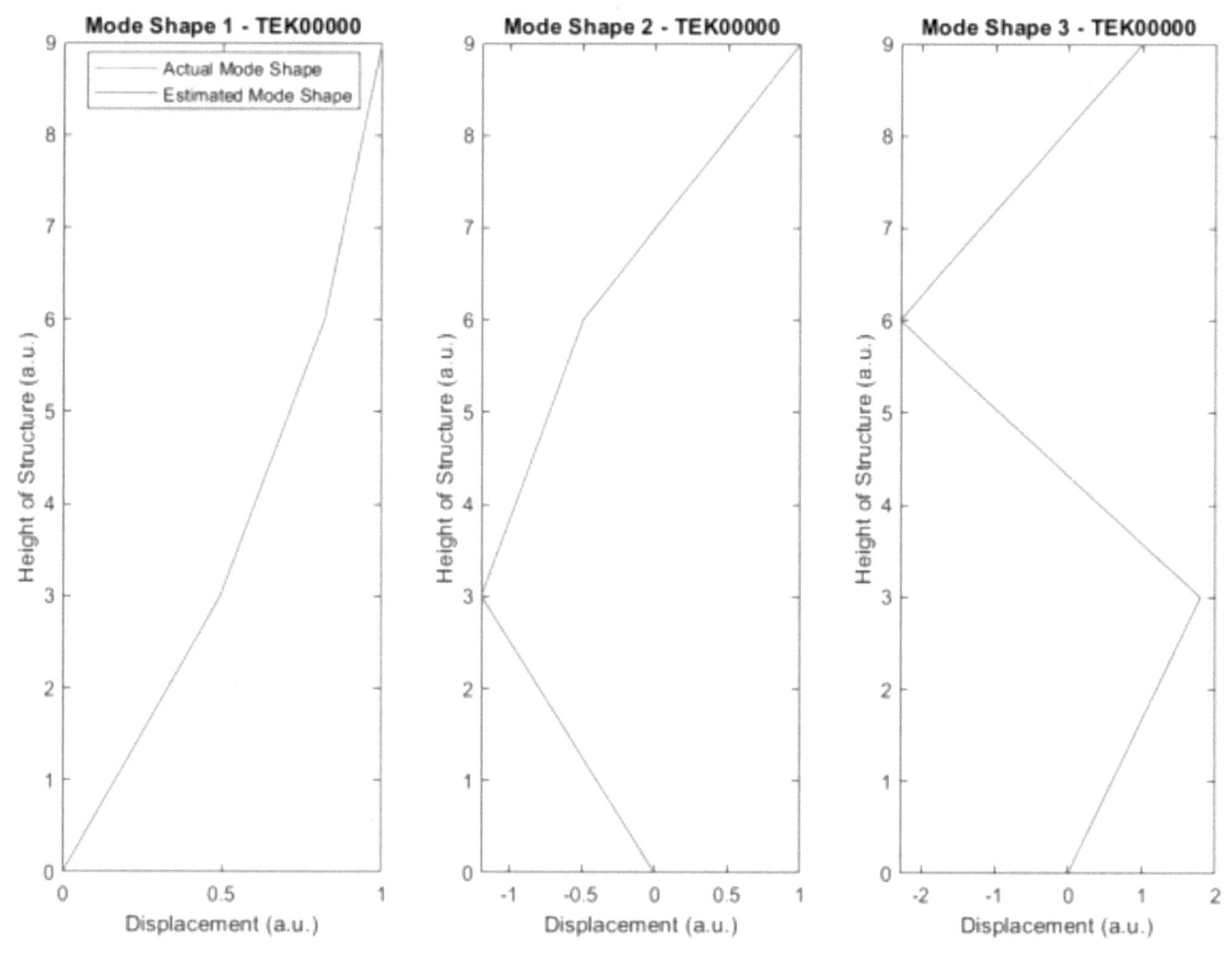

Fig. 3.5 Mode shapes of physical structure experiment using revised Sanger's learning rule

Fig 3.6a Conventional technique

0.999	0.001	0.001
0.001	0.999	0.001
0.001	0.001	0.999

Fig 3.6b Sanger's rule

0.999	0.001	0.001
0.001	0.999	0.001
0.001	0.001	0.999

Fig 3.6c Conventional versus analytical

0.999	0.001	0.001
0.001	0.999	0.001
0.001	0.001	0.999

Fig 3.6d Conventional versus Sanger's rule

0.999	0.001	0.001
0.001	0.999	0.001
0.001	0.001	0.999

References

1. Dorn, C., Dasari, S., Yang, Y., Farrar, C., Kenyon, G., Welch, P., et al.: Efficient full-field vibration measurements and operational modal analysis using neuromorphic event-based imaging. J. Eng. Mech. **144**(7), 04018054, 1–12 (2018)
2. Feng, D., Feng, M.Q.: Computer vision for SHM of civil infrastructure: from dynamic response measurement to damage detection – a review. Eng. Struct. **156**, 105–117 (2018)
3. Serov, A.: Cognitive sensor technology for structural health monitoring. Procedia Struct. Integr. **5**, 1160–1167 (2017)
4. Worden, K., Graeme, M.: The application of machine learning to structural health monitoring. Philos. Trans. R. Soc. **365**(1851), 515–537 (2007)
5. Rafiei, M.H., Adeli, H.: A novel machine learning-based algorithm to detect damage in high-rise building structures. Struct. Des. Tall Spec. Build. **26**(18), e1400, 1–11 (2017)
6. Salehi, H., Burgueno, R.: Emerging artificial intelligence methods in structural engineering. Eng. Struct. **171**, 170–189 (2018)
7. Poncelot, F., Kerschen, J., Golinval, C., Verhelst, D.: Output-only modal analysis using blind source separation techniques. Mech. Syst. Signal Process. **21**(6), 2335–2358 (2007)
8. Yang, Y., Nagarajaiah, S.: Blind modal identification of output-only structures in time-domain based on complexity pursuit. Earthq. Eng. Struct. Dyn. **42**(13), 1885–1905 (2013)
9. Yang, Y., Dorn, C., Mancini, T., Talken, Z., Kenyon, G., Farrar, C., et al.: Blind identification of full-field vibration modes from video measurements with phase-based video motion magnification. Mech. Syst. Signal Process. **85**, 567–590 (2017)
10. Yang, Y., Dorn, C., Mancini, T., Talken, Z., Theiler, J., Kenyon, G., et al.: Reference-free detection of minute, non-visible, damage using full-field high-resolution mode shapes output-only identified from digital videos of structures. Struct. Health Monit. **17**(3), 514–531 (2018)
11. Stone, J.V.: Independent Component Analysis: A Tutorial Introduction. A Bradford Book, Cambridge (2004)
12. Roy, K., Jaiswal, A., Panda, P.: Towards spike-based machine intelligence with neuromorphic computing. Nature. **575**(7784), 607–617 (2019)
13. Shastri, B., Tait, A.N., Ferreira de Lima, T., Pernice, W.H., Bhaskaran, H., Wright, C.D., et al.: Photonics for artificial intelligence and neuromorphic computing. Nature. **15**(2), 102–114 (2021)
14. Brunton, S.L., Kutz, J.: Data-Driven Science and Engineering: Machine Learning, Dynamical Systems, and Control. Cambridge University Press, Cambridge (2019)
15. Hyvarinen, A., Oja, E.: Independent component analysis by general nonlinear Hebbian-like learning rules. Signal Process. **64**, 301–313 (1998)
16. Jangid, R.S.: Response Analysis for Multi Support Earthquake Excitation. Department of Civil Engineering, Indian Institute of Technology Bombay, Mumbai (2013)
17. Lindfield, G., Penny, J.: Numerical Methods Using MATLAB, 4th edn. Academic Press (2019)
18. Yang, J.-Q., Wang, R., Yang, J.-Q., Wang, R., Han, S.-T., Ren, Y., Mao, J.-Y., Wang, Z.-P., Zhou, Y.: Neuromorphic engineering: from biological to spike-based hardware nervous systems. Adv. Mater. **32**(52), 2003610, 1–32 (2020)
19. Bao, Y., Li, H.: Machine learning paradigm for structural health monitoring. Struct. Health Monit. **20**(4), 1353–1372 (2020)
20. Watkins, Y., Thresher, A., Schultz, P., Wild, A., Sornborger, A., Kim, E., et al.: Towards self-organizing neuromorphic processors: unsupervised dictionary learning via a spiking locally competitive algorithm. International conference on neuromorphic systems (2019).

Chapter 4
Environmental-Insensitive Damage Features Based on Transmissibility Coherence

Samuel M. Hinerman, Makayla E. Ley, Peter A. Newman, James Gibert, Garrison S. Flynn, and Charles R. Farrar

Abstract Online damage identification is crucial to ensure safety and reliability of a wide array of civil, mechanical, and aerospace systems. In structural health monitoring (SHM), features are extracted from sensor data acquired on in-service structures to detect the presence of damage. These features are often modal parameters (e.g., natural frequencies, mode shapes) that can be sensitive to the damage and will indicate a damaged state when they deviate far enough away from a nominal condition. Environmental and operational variability (EOV) poses a significant challenge in determining the health state of a structure, because the EOV effects (e.g., temperature fluctuations) can produce changes in features that are similar to those produced by damage. Damage and EOV both affect the properties of structures (mass, stiffness, & damping). Structures are often designed to operate as a linear system, even under extreme operational and environmental loading, while damage (e.g., cracks, plastic deformation) can introduce nonlinearities into structures. In this study, transmissibility coherences (TC) are explored as potential features that are insensitive to EOV while remaining sensitive to damage. TC measures the linearity of output-output relationships across the frequency domain, collecting more spatial information and negating the need to measure inputs to the system when compared to traditional input-output coherence. Multiple TC-based metrics were defined and used to assess the changes in TC caused by damage. These metrics were validated with data from numerical and experimental tests of a 4-story structure and highway bridge structure, using an analysis of variance (ANOVA) sensitivity analysis to ensure that damage-sensitive features were sensitive to damage and insensitive to EOV. In this process F-tests were used to classify system response as healthy or damaged. The mean absolute deviation (MAD) of the TC performed well in experimental tests, yielding accuracy, precision, recall, and F1 scores of over 90%. Receiver operating characteristic (ROC) curves were employed to identify the optimal metric. When excluding low damage cases and data containing unwanted nonlinear influence in the undamaged state, the MAD of TC behaved as a perfect classifier.

Keywords Damage detection · Structural health monitoring · Coherence · Vibration testing

S. M. Hinerman
Department of Mechanical Engineering, University of California, Riverside, CA, USA
e-mail: shine003@ucr.edu

M. E. Ley
Department of Mechanical & Aerospace Engineering, New Mexico State University, Las Cruces, NM, USA
e-mail: makley@nmsu.edu

P. A. Newman
Department of Mechanical Engineering, Washington State University Vancouver, Vancouver, WA, USA
e-mail: peter.newman@wsu.edu

J. Gibert
Department of Mechanical Engineering, Purdue University, West Lafayette, IN, USA
e-mail: jgibert@purdue.edu

G. S. Flynn · C. R. Farrar (✉)
Los Alamos National Laboratory, Los Alamos, NM, USA
e-mail: garrison@lanl.gov; farrar@lanl.gov

R. Madarshahian, F. Hemez (eds.), *Data Science in Engineering, Volume 9*, Conference Proceedings of the Society for Experimental Mechanics Series, https://doi.org/10.1007/978-3-031-04122-8_4

4.1 Introduction

Structural health monitoring (SHM) attempts to detect, locate, and quantify damage in engineered systems. Damage can be defined as changes to the material or geometric properties that adversely affect current or future system performance. SHM is critical for many engineered systems, particularly in the realms of civil, aerospace, and automotive industries. The SHM process can be broken up into four stages: (1) operational evaluation, (2) data acquisition, (3) feature extraction, and (4) feature discrimination [1]. The first stage, operational evaluation, defines the damage to be detected, and the responses to be measured are defined in the second stage. Damage-sensitive features will then be extracted from these data in the third stage. Once features are extracted, these data can be incorporated into a statistical analysis designed to recognize patterns that indicate damage (Stage 4). Lastly a prognosis can be made by combining assessments based on feature classification with probabilistic future loading models and damage evolution models to predict performance level variables (e.g., time until service is necessary, remaining lifespan).

The primary motive for SHM is to ensure the safety, reliability, efficiency, and sustainability of existing and newly designed systems. One major challenge seen when implementing SHM for real-world systems is EOV (e.g., temperature, wind, humidity, traffic loading). It has become clear that EOV can produce changes in structural properties and boundary conditions that are difficult to distinguish from the changes produced by damage [1]. Quantifying the effects of EOV in SHM typically requires extensive amounts of training data from difficult-to-measure sources of variability (e.g., nonuniform temperature distributions, traffic loading, offshore operation). In order to develop a robust SHM algorithm, this work focuses on uncovering features that differentiate damage from non-threatening environmental and operational effects. In-service systems experiencing EOV is one of the main challenges in the practical applications of SHM [1].

4.2 Background

Techniques that reduce EOV effects continue to be a research topic of interest. The majority of SHM approaches to separating EOV effects from damage effects on structures involve measuring how modal properties change from EOV then projecting out or modeling the EOV. The downside to these techniques is that they often require an extensive amount of data, additional instrumentation, sometimes knowledge of the input, the undamaged responses, the damaged responses, and the EOV responses on a structure.

A previous research approach by Erazo et al. [2] used a statistical analysis of Kalman filter residuals to yield damage features that were able to separate EOV effects from damage effects. This study numerically and experimentally devised a method that used a Bayesian whiteness test on the residual error of the Kalman filter as a damage-sensitive feature and then used a normalized damage index to measure the extent of the damage. The findings showed that this method was able to remain insensitive to varied uniform and nonuniform temperature distributions and was successful in the measurement and quantification of damage. For optimal results, this method required as much training data as possible on the healthy structure under daily temperature variations.

Similarly, Li et al. [3] applied Bayesian fusion for a damage identification strategy. This involved combining multiple data types to obtain a useful, accurate representation of a structure. Numerical studies yielded promising results for damage identification, but when experimental tests were conducted, false positives occurred. When the method was implemented for a real-world structure, false positives persisted, and the method was found to be ineffective for low-level damage cases. The authors note that issues may be due to the lack of damage information and conclude further studies are needed to devise an optimal combination of damage indices.

Using a machine learning algorithm, Huang et al. [4] was able to account for the effects of nonuniform temperature distributions in two different numerical models of multi-span bridges using modal parameters as damage features. This method reliably detected the presence of damage but was not able to consistently determine the location of the damage. This approach required extensive training on the environmentally varied undamaged structure but was ultimately effective in separating nonuniform temperature effects from damage.

Data normalization is commonly used to separate EOV effects from damage [5]. Normalization approaches establish a relationship between either input-output or output-only measurements. Operational measurements can then be observed in efforts to detect abnormalities that may arise when damage is present [5]. Output-only methods have distinct advantages for practical SHM as difficulties arise when attempting to capture the input excitation of in-operation structures.

Coherence can be used to measure the linearity between two signals and has been proposed as a damage index, because nonlinear behavior can be related to damage. Traditionally, coherence functions examine the input-to-output relationship of signals in a structure. Transmissibility coherence (TC) measures the linearity between two output signals and is advantageous for SHM applications because the input to the system is not required to be measured. Using single-input multiple-output (SIMO) numerical and experimental models, TC was studied in comparison to traditional coherence by Zhou et al. [6]. Damage indicators were incorporated by integrating the TC over the frequency bandwidth and calculating TC modal assurance criteria. Zhou et al. [7] further explores the feasibility of TC as a damage indicator by applying principal component analysis enhanced with distance measure. Results show that coherence has a promising future in SHM.

4.3 Motivation

Analyzing data from idealized systems usually tested in a well-controlled laboratory setting is a necessary first step for developing a SHM procedure. However, such an approach has led to many existing methods neglecting EOV effects [1]. EOV must be considered because these effects alter the dynamic response characteristics of a structure. The purpose of this paper is to derive a damage identification process that is sensitive to damage while insensitive to EOV. By incorporating TC as a damage-sensitive feature and using statistical models, changes in this feature caused by damage will be differentiated from changes caused by EOV effects. The proposed method will account for factors that can decrease coherence such as low signal-to-noise ratio, insufficient sample sizes, and leakage. The proposed method will by demonstrated with numerical and experimental data from a single-input, multiple-output (SIMO) 4-story structure and with experimental data obtained from an in *in situ* highway bridge.

4.4 Methodology

This study focuses on developing a statistical model for damage detection which is insensitive to multiple sources of time-varying EOVs. In addition, the method requires output-only responses of a structure, making it more feasible for deployment in real-world scenarios. It is hypothesized that TC is insensitive to linear changes caused by EOV that systems encounter. Additionally, when damage presents a nonlinear response, coherence will decrease in a detectable manner. The underlying assumptions are as follows: (1) the structure can be accurately represented as a linear system in its undamaged state; (2) while EOV may cause changes in structural properties (mass, stiffness, damping), the system remains linear; and (3) damage will produce an observable nonlinear response as indicated by a reduction in TC.

4.5 Transmissibility Coherence

Coherence is commonly used in vibration testing as a tool to assess whether an experiment is well conducted [1]. Traditionally, coherence measures the strength of the linear relationship between an input and an output. In this study, this measurement of linearity is used in combination with healthy and damaged acceleration time histories of structures receiving random excitation to develop a damage detection methodology that is quick and easy to implement. TC is defined in Eq. 4.1 where G_{y1} and G_{y2} are the output power spectrums, and $G_{y1,y2}$ is the cross-power spectrum between the two outputs.

$$\mathrm{TC}(f) = \frac{\left|G_{y1,y2}(f)\right|^2}{G_{y1}(f)G_{y2}(f)} \tag{4.1}$$

TC is derived from the linear relationship between the input and various outputs of a structure in the frequency domain via the respective frequency response functions. The block diagram in Fig. 4.1 demonstrates this linear relationship for a SIMO structure. To derive a linear relationship between the two outputs, first note the linear relationships between the input power spectrum, G_{x1}, and the output power spectra that are defined by the respective frequency response functions, $H_{y1,x1}$ and $H_{y2,x1}$. By equating the values of G_{x1} in these two relationships, one arrives at the relationship that shows $G_{y1}$and G_{y2} are linearly related by the ratio of the magnitude-squared FRFs. As with the traditional coherence, the TC provides a measure of the linear relationship between output measures. Figure 4.2 shows examples of the traditional input-output coherence and the TC obtained from measurements on the 4-story structure described below when the structure was in a baseline undamaged state (green), an undamaged state with simulated EOV (blue) and a damage state (red) where the damage introduces a

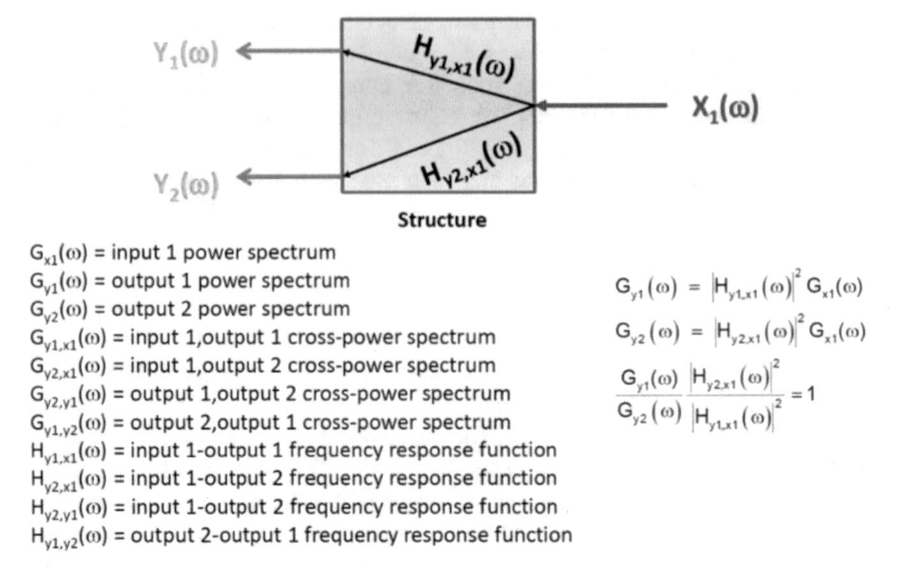

Fig. 4.1 Block diagram of linear relationship between SIMO structures in the frequency domain

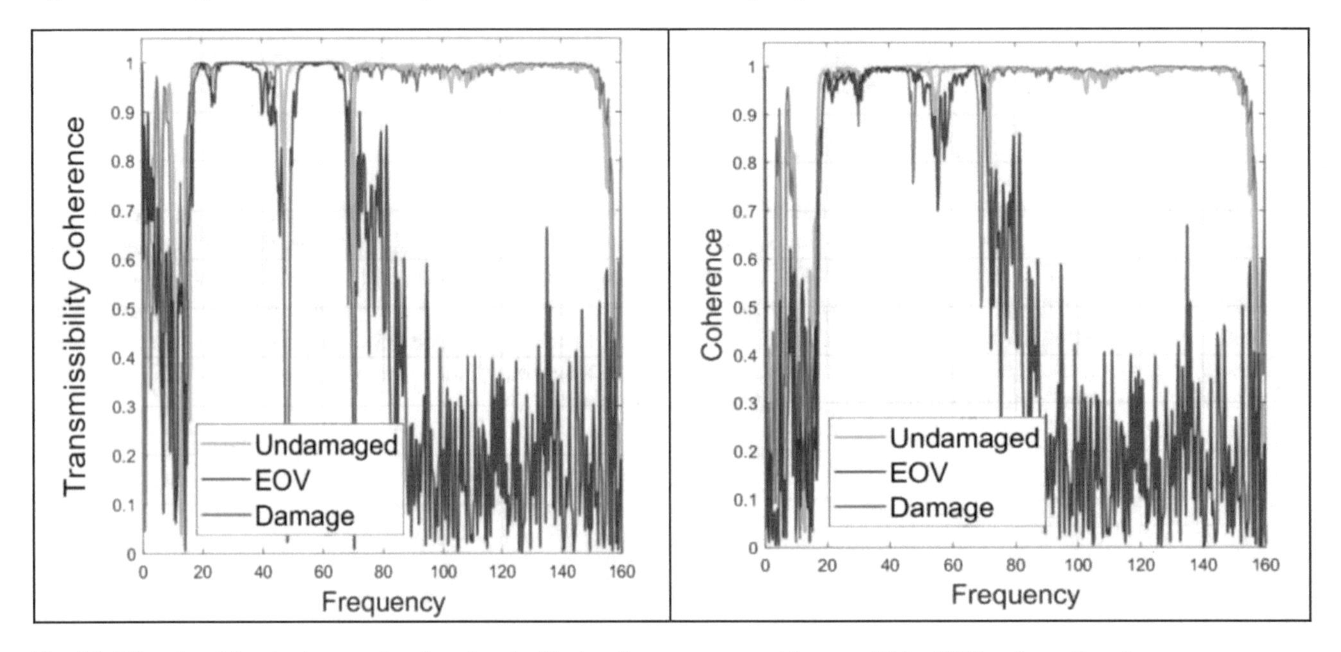

Fig. 4.2 TC and traditional coherence functions for a healthy baseline case, a case with mass addition EOV, and a medium damage case

nonlinearity to the system. Note that the structure was subjected to a 20–150 Hz band-limited white-noise input for these measurements.

4.6 Statistical Features

Statistical features are applied to TC functions in order to condense the TC functions into scalar features that can be compared with an F-test. A wide variety of features, including but not limited to the mean, variance, area under the curve, and mean absolute deviation (see equations below), are computed. An ideal feature show that the TC function did not change when EOV is affecting the undamaged system and that the TC changed when damage that produces a nonlinear response is affecting the structure. In this study, multiple statistical features were applied to the TC functions however the metrics below are focused on.

For the purpose of damage detection, a statistical metric is applied to TC functions, reducing the function to scalar values. We define the mean (Eq. 4.2) area under the curve (AUC) (Eq. 4.3), mean absolute deviation (MAD) (Eq. 4.4), and variance (Eq. 4.5) as statistical features of the TC.

$$\frac{\sum_{i=1}^{n} x_i}{n} \tag{4.2}$$

$$\int_{\omega_{\min}}^{\omega_{\max}} x_i \tag{4.3}$$

$$\frac{1}{n}\sum_{i=1}^{n} |x_i - \overline{x}| \tag{4.4}$$

$$\frac{\sum_{i=1}^{n} (x_i - \overline{x})^2}{n-1} \tag{4.5}$$

4.7 Damage State Classification: F-Test

A statistical hypothesis test (F-test) is implemented as a tool for damage detection. For an ideal feature, the hypothesis test would show that healthy datasets without EOV and healthy datasets with EOV are similar, and damaged data sets are significantly different from the prior two. The F-test evaluates the variation of a feature within groups (Eq. 4.6) relative to the variance between groups (Eq. 4.7). In these equations, *MG* is the mean value of a group of data, *MT* is the mean of values from all groups, *n* is the number of samples in the group, and *m* is the number of groups. Finally, the F-statistic, which is the ratio of the variance between and within groups, is calculated and compared to an F-critical value. F-critical is determined using a lookup table based on a desired confidence value. It is based upon the denominators of *VW* and *VB,* which are known as the degrees of freedom. F-critical varies depending on the significance level which is inversely proportional to the confidence level. This relationship means that when the significance level decreases, the F-critical value increases. A higher F-critical provides less of a risk in concluding that there is a difference in groups when in actuality there is not an outstanding difference. If the F-statistic exceeds F-critical, the null hypothesis is rejected and the alternative hypothesis is accepted.

$$VW = \frac{\sum_{i=1}^{n} (y_i - MG)}{n\,(m-1)} \tag{4.6}$$

$$VB = \frac{\sum_{i=1}^{n} (MG - MT)}{(n-1)} \tag{4.7}$$

$$\mathrm{F-statistic} = \frac{VB}{VW} \tag{4.8}$$

In this study, an F-test is used to classify sets of data as either damaged or healthy. This process is similar to logistic regression in that both involve predicting whether something is true or false. We predict whether the state of the structure is healthy or damaged. For this F-test, the null hypothesis is that the structure is healthy and the alternative hypothesis is that the structure is damaged. If the variance between groups is significantly larger than the variance within groups, the null hypothesis is rejected, and the system is identified as damaged. An example of an ideal set of data is shown in Fig. 4.3 where the mean of transmissibility coherence functions was calculated and separated into three classes: a healthy baseline class, a healthy class exposed to mass EOV, and a medium-damaged class. As seen, there is little variance within the healthy class, meaning the methodology applied to these data is likely insensitive to the effects of EOV. When comparing the healthy classes to the damaged, there is detectable deviation.

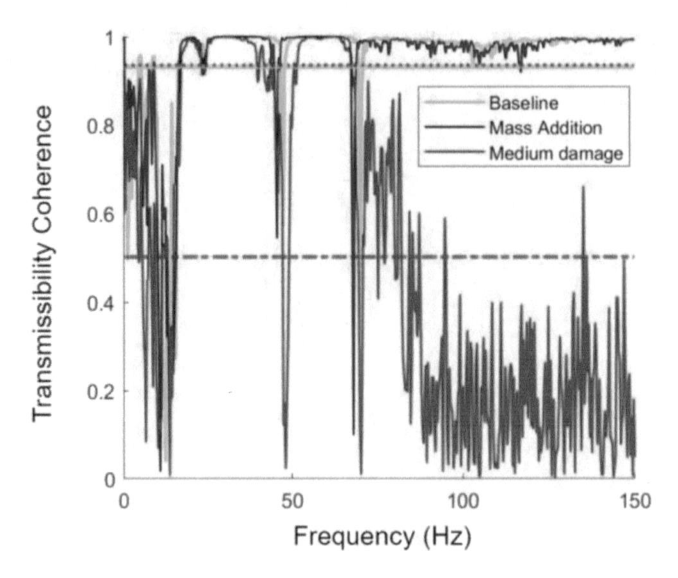

Fig. 4.3 A portion of a set of data displaying the TC functions and the mean of these TCs

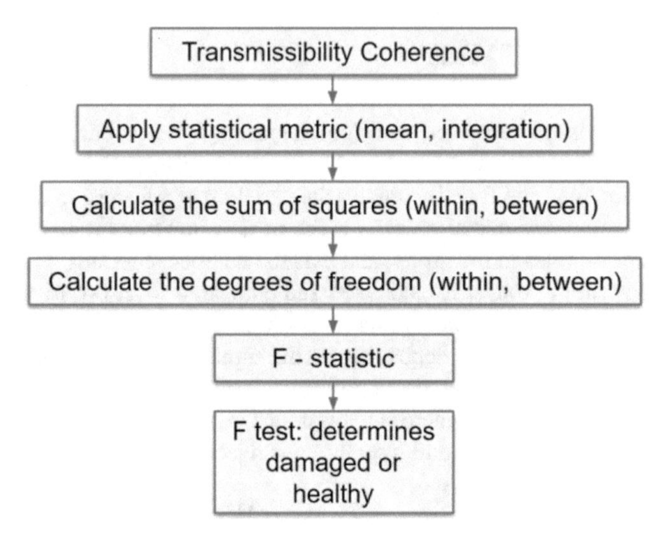

Fig. 4.4 Classification flowchart

The procedure followed in this damage detection algorithm is displayed in Fig. 4.4. First the transmissibility coherence is extracted from two outputs. Next, a statistical metric is applied to the function. Then, using a collection of statistical metrics extracted from TC, the sum of squares and degrees of freedom, within and between, is calculated. This allows for the calculation of an F-statistic. An F-test entails a comparison of F-statistic to F-critical. If the F-statistic is greater than F-critical, the system is classified as damaged.

We characterize performance of the classifier by taking note of the number of false positives (damage detected when it is not present) and false negatives (healthy detected when the structure is damaged). A classification evaluation is conducted using receiver operating characteristic (ROC) curves, accuracy, precision, recall, and F1 scores. This process allows for quantification of how well the various statistical metrics perform, for the purpose of choosing the highest performing metric.

Determination of the optimal statistical metric to extract from TC, the various classification models must be evaluated. This is done by first applying the algorithm to a data set with known health states. The predictions of the model are compared to the actual state, and the number of true positives (TP), true negatives (TN), false negatives (FN), and false positives (FP) is tracked. Performance evaluators such as accuracy, precision, recall, F1 scores, and receiver operating characteristic (ROC) curves can then be determined.

4.8 Sensitivity Analysis: Main Effect Screening by Analysis of Variance

Sensitivity analysis is performed on computational models. An effect screening of the numerically modeled 4-story structure is performed by using analysis of variance (ANOVA). Using this technique, the relationship between response variables and independent variables can be investigated and modeled. This method requires a full factorial design of experiments (DOE). The key steps followed to design an experiment are the following: (1) identify factors of interest and response variables, (2) determine appropriate levels for each explanatory variable, (3) determine a design structure, (4) randomize the order in which each set of conditions in ran and then collect the data, and (5) organize the results. At this point, appropriate conclusions are drawn by observing the effects of varying the levels of the factors of interest. By doing this, the effects of EOV (e.g., mass and stiffness variation) are compared to damage. The goal is to determine the statistical metrics that are sensitive to damage while remaining insensitive to EOV.

In this study, sensitivity analysis is a formal process for determining which factors have an influence on the statistics extracted from TC and hence on the classification process. Sensitivity analysis is implemented through effect screening with an end goal of choosing the statistical metrics best suited for classification. The characteristics strived for are sensitivity to damage and insensitivity to EOV. The effects of stiffness reduction, mass addition, and crack simulation were examined by calculating the R-squared percentage of a numerical full factorial DOE dataset. The R-squared percentage is based on the residual sum of squares (RSS) which measures the variation a factor introduces as well as the total sum of squares (TSS) which provides a measurement of the variation among all experiments.

$$R^2 = 1 - \frac{\text{RSS}}{\text{TSS}} \tag{4.9}$$

By calculating the R-squared percentage, the amount of variance a dependent variable experiences due to an independent variable is quantified. The analysis of results provides an understanding of the effects that stiffness changes which approximate thermal effects, mass changes which approximate operational variability, and the system health state have on features used to compare changes in TC.

4.9 Test Structure

The 4-story structure used for experimental tests and as a basis for numerical modeling is shown below in Fig. 4.5. As pictured, the structure is mounted on linear rails that allowed horizontal motion in one direction, and a single shaker excites the base in that same direction. The input force is measured by a force transducer on the tip of the shaker, and the structural response is measured by the blue accelerometers on each floor. Damage is simulated by setting a small gap between the bumper and column mechanism in between the top two floors so that the bumper and column contact when the system is excited. This mechanism introduces a nonlinear change in the stiffness between those floors and can be compared to changes in stiffness that might result from a crack opening and closing under similar conditions. In both the experimental and numerical cases, only the Single-Input Multiple-Output (SIMO) case is considered, with a single excitation delivered to the base. The drawing of the structure in Fig. 4.5 details the structure further and labels the accelerometers with their appropriate channels.

4.10 Experimental Tests

To test the validity of the proposed damage detection algorithm, time histories of acceleration responses were collected for the experimental 4-story structure. Sets of fifty time histories for the four accelerometers on the structure were recorded for each of the 17 different structural states (50x17=850 total). The 17 states include a healthy baseline, varying mass and stiffness EOVs, as well as simulated damage conditions, detailed in Table 4.1 below. In this case, mass changes can represent operational variability (e.g., varying live loads on a structure), and local stiffness changes can represent the effects of nonuniform temperature distributions across the structure.

The severity of the simulated damage increases as the gap distance decreases because the bumper and column will impact more frequently and with more energy as the distance between them is reduced. To limit any extraneous noise entering through the base of the setup, the experimental structure and shaker were placed on top of a 12.5-cm rigid foam sheet.

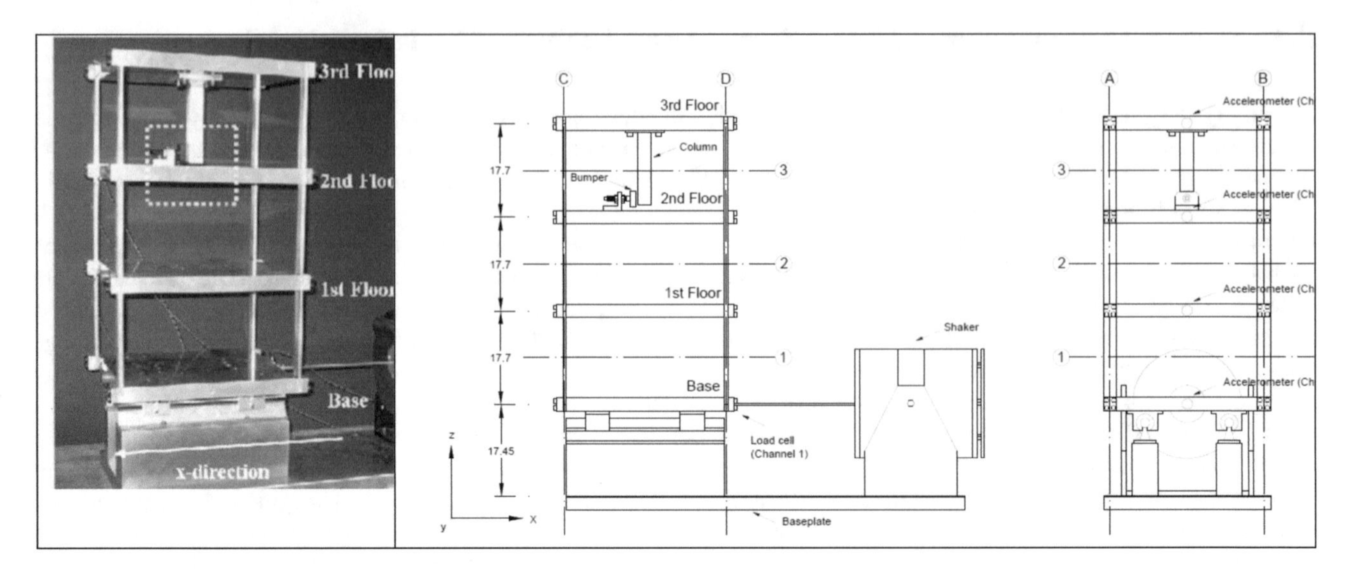

Fig. 4.5 The 4-story experimental structure and related drawing

Table 4.1 Structural states for the experimental tests

State	Description
1	Baseline condition
2	Mass = 1.2 kg at base
3	Mass = 1.2 kg at 1st floor
4	87.5% stiffness reduction in column 1BD
5	87.5% stiffness reduction in column 1AD and 1BD
6	87.5% stiffness reduction in column 2BD
7	87.5% stiffness reduction in column 2AD and 2 BD
8	87.5% stiffness reduction in column 3BD
9	87.5% stiffness reduction in column 3AD and 3BD
10	Gap = 0.20 mm
11	Gap = 0.15 mm
12	Gap = 0.13 mm
13	Gap = 0.10 mm
14	Gap = 0.05 mm
15	Gap = 0.20 mm and mass = 1.2 kg at base
16	Gap = 0.20 mm and mass = 1.2 kg at 1st floor
17	Gap = 0.10 mm and mass = 1.2 kg at 1st floor

However, a bolt on the 2nd floor (output 4) was discovered to be loose after the experiment had been conducted. The bolt not being secured properly was believed to have introduced nonlinearities into the output 4 response, which violates one of the base assumptions that the undamaged structure is a linear system. For the results of the experimental tests, some cases will be considered excluding all data related to output 4.

4.11 Numerical Model – 4-Story Structure

A lumped mass model of the 4-story structure is used to numerically validate damage features alongside the experimental tests (Fig. 4.6). A validation study by Nishio et al. [8] regarding the modeling of this specific structure was referenced for the setup of the numerical model.

The model represents the equivalent masses of all the components in each floor as point masses. Each set of columns are assumed to be fixed in rotation and act as springs/dampers in parallel, allowing the stiffness and damping of the columns between each floor to be consolidated. The stiffness of each individual column can be calculated using Euler-Bernoulli beam theory ($12EI/L^3$), where E is the modulus of elasticity (65 GPa, aluminum), I is the cross-sectional area moment of inertia,

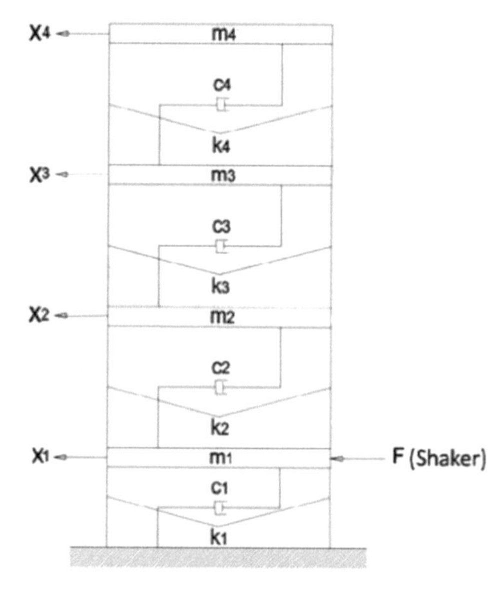

Fig. 4.6 Lumped mass equivalent of the 4-story structure

Table 4.2 Baseline model parameters

Floor	Mass (kg)	Stiffness (N/m)	Modal damping ratio
1	6.54	10	0.01
2	6.66	399790	0.01
3	6.86	399790	0.01
4	6.8	399790	0.01

and L is the length of the column. The baseline values of the model parameters are listed in Table 4.2. The stiffness of the first floor is used to approximate the resistance between the base floor and the rails it is mounted on.

The specific matrices generated from the model parameters are given below.

$$[M] = \begin{bmatrix} m1 & 0 & 0 & 0 \\ 0 & m2 & 0 & 0 \\ 0 & 0 & m3 & 0 \\ 0 & 0 & 0 & m4 \end{bmatrix} \tag{4.10}$$

$$[K] = \begin{bmatrix} k_1 + k_2 & -k_2 & 0 & 0 \\ -k_2 & k_2 + k_3 & -k_3 & 0 \\ 0 & -k_3 & k_3 + k_4 & -k_4 \\ 0 & 0 & -k_4 & k_4 \end{bmatrix} \tag{4.11}$$

$$\left[C^M\right] = \begin{bmatrix} 2\zeta_1\omega_1 M_1 & 0 & 0 & 0 \\ 0 & 2\zeta_2\omega_2 M_2 & 0 & 0 \\ 0 & 0 & 2\zeta_3\omega_3 M_3 & 0 \\ 0 & 0 & 0 & 2\zeta_4\omega_4 M_4 \end{bmatrix} \tag{4.12}$$

The modal damping matrix $[C^M]$ is transformed into the damping matrix using the mode shape matrix $[\phi]$.

$$[C] = \left[[\phi]^T\right]^{-1}\left[C^M\right][\phi]^{-1} \tag{4.13}$$

There were issues in the model outputs documented in Nishio's report attributed to the interpolation scheme within the MATLAB solver, resulting in noisy high-frequency response characteristics. This effect appeared to be removed by setting the maximum step of the ODE45 solver to 0.1 times the period of the maximum frequency of interest, as suggested in ASCE Standard 4–16 [9] regarding implicit linear response-history solvers.

4.12 Modeling of Damage and EOV

The damage simulated in the model is intended to reflect the mechanism from the experimental structure detailed in Fig. 4.7. When the displacement between the top two floors exceeds the gap between the bumper and center column, the stiffness between those floors significantly increases.

Rather than assuming the behavior of an ideal bilinear spring, a smooth delta (Δ) region is implemented to transition the stiffness from one state to the next while maintaining C1-continuity. The equations given for modeling damage in the report by Nishio do not match the intended behavior, shown in Fig. 4.8. Below is a modification of those equations that maintain the C1-continuity of the force vs. displacement curve.

$$X \leq \text{gap} : k_4 = k_c \tag{4.14}$$

$$\text{gap} < X < \text{gap} + \Delta : k_4 = k_c - k_b * \frac{\text{gap}}{\Delta} + \frac{k_b * X}{2\Delta} + \frac{k_b * \text{gap}^2}{2\Delta * \text{X}} \tag{4.15}$$

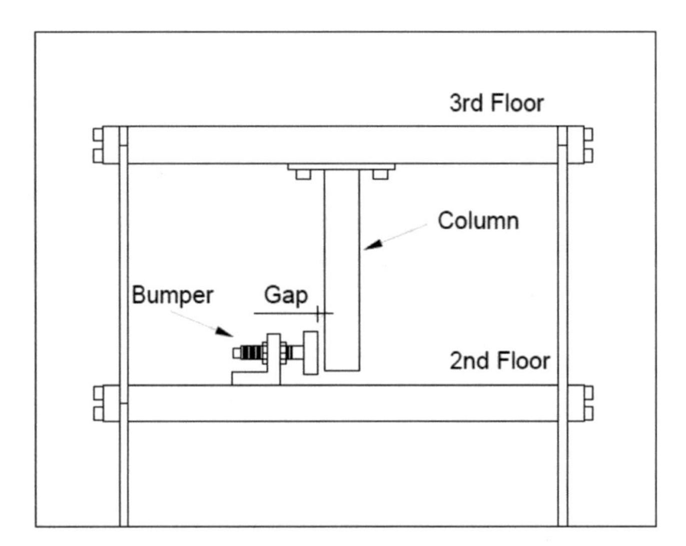

Fig. 4.7 Mechanism simulating damage in the 4-story structure

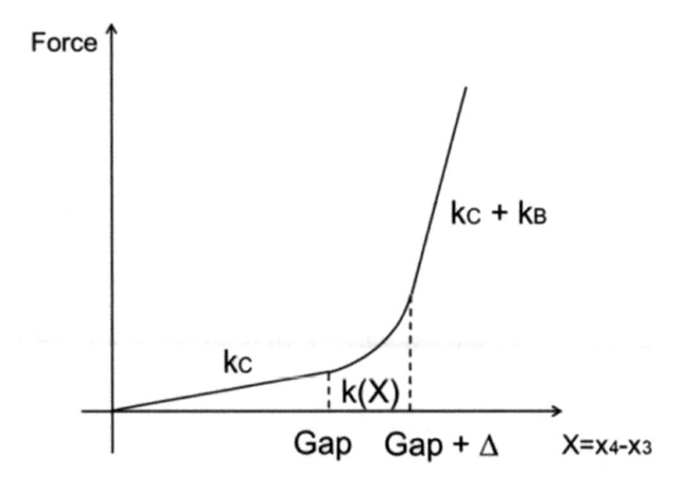

Fig. 4.8 Force-displacement curve for the relative displacement between the top 2 floors. K_C is the original column stiffness, and K_B is the stiffness of the central column

$$X \geq \text{gap} + \Delta : k_4 = k_c + k_b - k_b * \frac{\left(\text{gap} + \frac{\Delta}{2}\right)}{\text{X}} \tag{4.16}$$

The validation study from Nishio yielded an optimized selection of Δ based on the gap distance that would most accurately reflect the experimental behavior, but because the equations were modified, those results are not applicable. Additionally, the magnitude of the response of the model exceeded the experimental results, so the gap distance was scaled up to match the increased response. The EOV simulated in the model is similar to the experimental structure. Local stiffness reductions between floors attempt to simulate nonuniform temperature effects, and local mass additions attempt to simulate operational loading effects.

4.13 Results and Discussion

A damage detection algorithm was developed and its robustness was investigated. Before applying the method to experimental data, a sensitivity analysis was implemented using numerical data. The analysis of a full factorial DOE resulted in an understanding of the effects that stiffness changes, mass changes, and the system health state have on features used to compare changes in TC. Once the TC features were confirmed to demonstrate sensitivity to damage and insensitivity to EOV, system complexity and uncertainties were increased in the analysis of experimental data. Lastly, real-word partial validation was performed by demonstrating that the classification methodology is insensitive to real-world EOV effects.

4.14 Damage State Classification: Experimental Data

As mentioned, an F-test is used to classify sets of data as either damaged or healthy. The experimental set of data from the 4-story structure which has known health states was used as training data. The F-test was applied to the dataset, and the classification F-statistic results were tabulated. In order to find the most effective statistical feature, receiver operator characteristic (ROC) curves were plotted as shown in Figs. 4.9 and 4.10. ROC curves were produced by varying the significance level that, in turn, varies the threshold for damage detection.

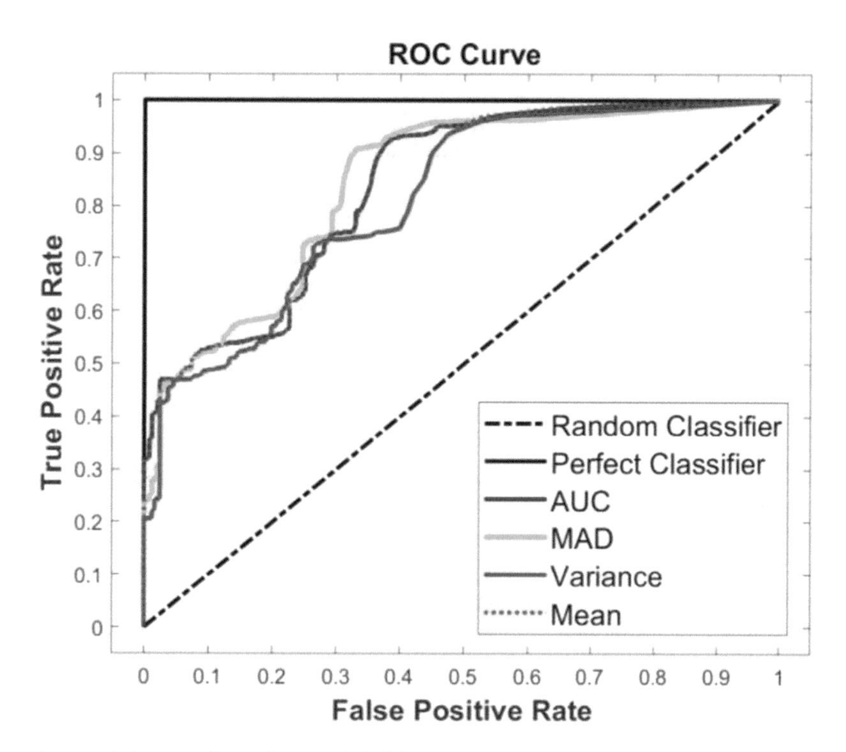

Fig. 4.9 Receiver operator characteristic curve for various statistical features

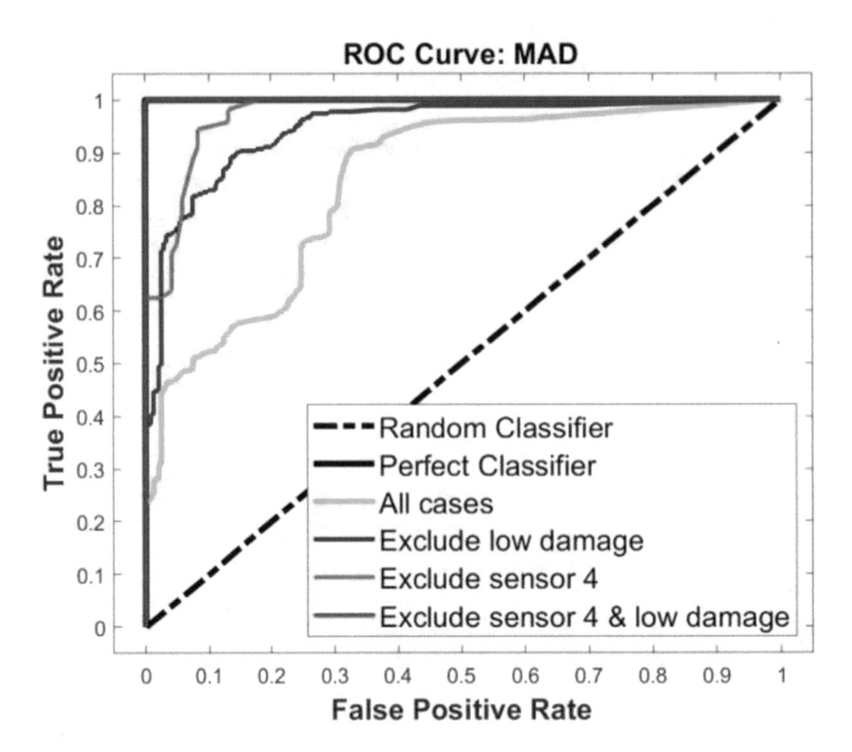

Fig. 4.10 Receiver operator characteristic curve for different data ranges of mean absolute deviation

As the y axis (TPR) increases, the amount of false-negative errors decreases. As the x axis (FPR) increases, the amount of false-positive errors increases. As a result, a perfect classifier would fall in the top-left corner of the graph where the TPR is one, the FPR is zero, and no errors are present. An ideal classifier that this study is searching for will maximize its area under the curve and be as close to this perfect classifier as possible. While all the features perform well, the MAD is a statistical feature that performs the best and is analyzed further.

In order to further investigate MAD of TC as a damage-sensitive feature, the data used in the F-test was limited. A violation of the hypothesis was found within the sensor 4 data. These data consisted of an unwanted nonlinear response from an unsecure bolt. When excluding sensor 4 data, the classification accuracy improved drastically. Additionally, it was noticed that the F-statistics for all the low-damage cases produced a noticeable amount of false-negative errors when conducting the F-test, so low-damage data was excluded. The result was the MAD ROC curve moving closer to behaving as a perfect classifier. Excluding both the low damage and sensor four cases yielded a ROC curve showing a perfect classifier.

4.15 Sensitivity Analysis: Numerical Model Data

A two-level factorial DOE was applied to the model to understand better how the modeled EOV and damage affects the model (Table 4.3). EOV is simulated in the model in the same manner as the experimental structure. Local stiffness reductions between floors attempt to simulate nonuniform temperature effects, and local mass additions attempt to simulate operational effects. Two levels of stiffness were introduced, a baseline as well as a 12.5% local stiffness reduction, and two levels of mass were introduced, the baseline as well as a mass addition of 1.2 kg on the second floor. Two levels of damage were used, healthy and high damage, based on the scaled gap values for the model. Model setup, run-time, and data analysis prevented a more comprehensive DOE, in which the EOV would be applied to more than one floor, at more than two levels of severity.

Sensitivity screening of data from the numerical model yielded reassuring results. Figure 4.11 shows the effect of the stiffness, mass, and damage on the MAD and mean metrics for each sensor pair. Damage showed the highest sensitivity for almost all pairings but was less sensitive between sensor pairs farther from the simulated damage, shown by pair 23. Inversely, the sensor pair closest to the damage (pair 45) was the most sensitive to damage and had almost no effects from the EOV simulated near the bottom of the structure. These results show that the method is sensitive to damage location. These correlations show promising results for future work regarding the identification of the location of damage using this method.

Table 4.3 Sensitivity full-factorial design of experiments (DOE)

Run #	Stiffness of k_2 (N/m)	Added mass to m_2 (kg)	Damage state
1	399790	0	Healthy
2	349816	0	Healthy
3	399790	1.2	Healthy
4	349816	1.2	Healthy
5	399790	0	Damaged
6	349816	0	Damaged
7	399790	1.2	Damaged
8	349816	1.2	Damaged

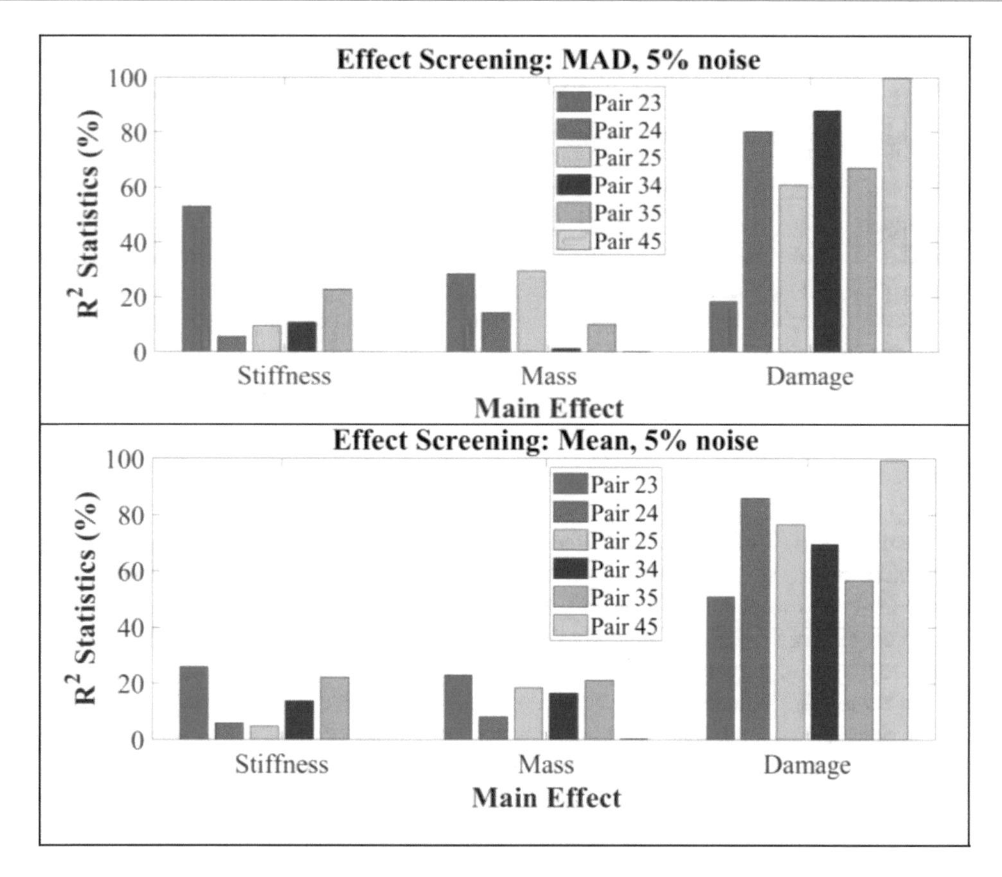

Fig. 4.11 Results of effect screening with 5% noise added to signals for (top) MAD of the TC and (bottom) mean of the TC

4.16 Real-World Example: Alamosa Canyon Bridge

The Alamosa Canyon Bridge (ACB) data used was a 24-hour impact test where eleven individual impact tests were made during a 1-day period. Each of the 11 impact tests involved 30 individual impacts on the same driving point on the bridge (driving point A) over a 2-hour period [10]. The data from ACB represented an undamaged in situ structure that was exposed to varying EOV conditions mainly sourcing from temperature fluctuation. Therefore, the data was used to partially validate transmissibility coherence as a damage detection feature that is insensitive to EOV.

The last study presented demonstrates that this methodology is insensitive to real-world EOV effects. Dynamic responses of the Alamosa Canyon Bridge, which experiences temperature variability, are analyzed using an F-test. More specifically, an F-test is applied to dynamic responses of the Alamosa Canyon Bridge – a healthy structure. In this experiment, an impact excitation was applied and data were collected at 11 different time intervals. For each time interval, there are 6 unique TC functions to which statistical metrics were applied. In order to compare an appropriate amount of data, 10 out of the 11 time intervals are randomly selected and divided into 2 groups. The F-test is performed on the 2 groups, each having 30 scalar values, and the F-statistic is recorded. This process was repeated 100 times for accuracy.

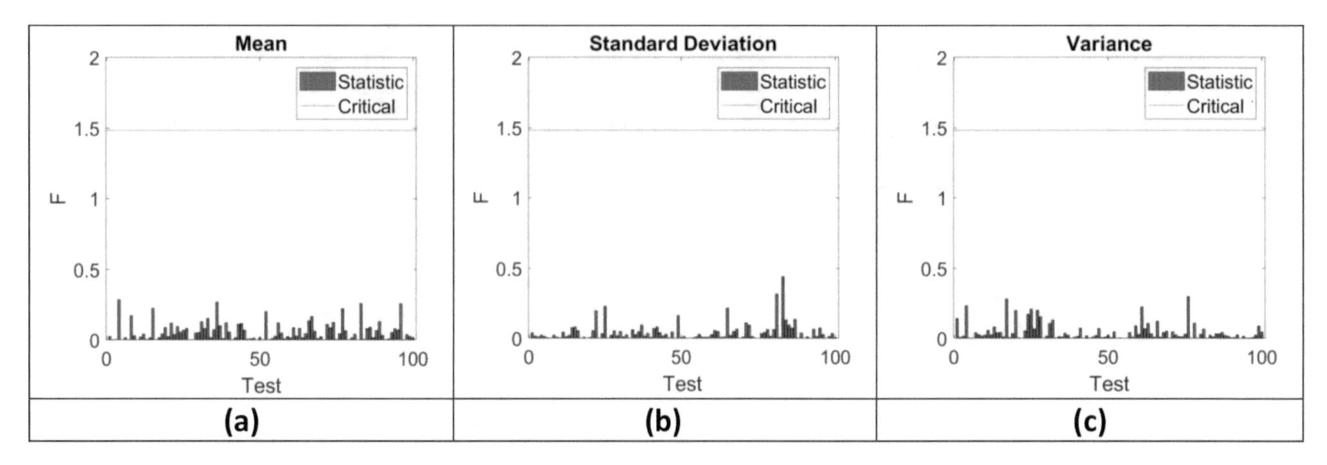

Fig. 4.12 Alamosa Canyon Bridge F-statistic using the following TC statistical metrics: (**a**) mean, (**b**) standard deviation, and (**c**) variance

The results are displayed in Fig. 4.12. The three statistical features of the TC functions that the F-test was used to evaluate each showed values all well below the F-critical value, indicating no cases of damage for the Alamosa Canyon Bridge. Overall, the data from the Alamosa Canyon Bridge was useful in confirming that applying statistical features to the TC functions would yield no false-positive errors. However, these data were not able to provide insight into this method's ability to avoid false-negative errors.

4.17 Conclusion

A method that is insensitive to EOV effects and sensitive to damage is important to achieve for SHM applications. In this study, determining whether a structure is healthy or damaged is a critical task efficiently completed through use of an F-test. Experimental and analytic structures were investigated, and it was found that the statistical features extracted from transmissibility coherence performed well in being sensitive to damage and insensitive to EOV. Nonlinearities associated with sensor 4 data violate the assumptions in the hypothesis. Therefore, it was reassuring to find that classification improves when sensor 4 data are removed. Difficulties arise when the method attempts to detect low damage, for it leads to false-negative errors. For the data sets studied, a good classifier based on statistics of the TC was developed that improves even more when removing nonlinearities and focusing on medium to high damage. When applied to a real-world structure with EOV, TC features did not yield any false-positives. For practical SHM applications, a method that does not require an input is quick to calculate, and a method that uses minimal training data is ideal. In this study, the TC was shown to be a method that meets these criteria. Investigation of whether detectability improves depending on sensor location remains a topic of interest for future work as do cases when there are multiple inputs to the system.

Acknowledgments This research was funded by Los Alamos National Laboratory (LANL) through the Engineering Institute's Los Alamos Dynamics Summer School. The Engineering Institute is a research and education collaboration between LANL and the University of California San Diego's Jacobs School of Engineering. This collaboration seeks to promote multidisciplinary engineering research that develops and integrates advanced predictive modeling, novel sensing systems, and new developments in information technology to address LANL mission-relevant problems.

References

1. Farrar, C.R., Worden, K.: Structural Health Monitoring: A Machine Learning Approach. Wiley (2012)
2. Erazo, K., Nagarajaiah, S., Sen, D., Sun, L.: Vibration-based structural health monitoring under changing environmental conditions using Kalman filtering. Mech. Syst. Signal Process. **117**, 1–15 (2019)
3. Li, X.Y., Lin, S.J., Law, S.S., Lin, Y.Z., Lin, J.F.: Fusion of structural damage identification results from different test scenarios and evaluation indices in structural health monitoring. Struct. Health Monit. (2020). https://doi.org/10.1177/1475921720962168
4. Huang, M., et al.: Structural damage identification of steel-concrete composite bridge under temperature effects based on cuckoo search. Int. J. Lifecycle Perform. Eng. **3**(2), 111 (2019)

5. Tomé, E.S., Pimentel, M., Figueiras, J.: Damage detection under environmental and operational effects using cointegration analysis–application to experimental data from a cable-stayed bridge. Mech. Syst. Signal Process. **135**, 106386 (2020)
6. Zhou, Y.L., Figueiredo, E., Maia, N., Perera, R.: Damage Detection and Quantification Using Transmissibility Coherence Analysis, Shock Vib. 290714, 16, (2015). https://doi.org/10.1155/2015/290714
7. Zhou, Y.L., Maia, N.M., Abdel Wahab, M.: Damage detection using transmissibility compressed by principal component analysis enhanced with distance measure. J. Vib. Control. **24**(10), 2001–2019 (2018)
8. Farrar, C., Nishio, M., Hemez, F., Stull, C., Park, G., Cornwell, P., Figueiredo, E., Luscher, D.J., Worden, K.: Feature extraction for structural dynamics model validation. Los Alamos National Laboratory Report LA-14489, 2016
9. ASCE/SEI Standards 4-16, seismic analysis of safety-related nuclear structures, p 17
10. Farrar, C.R, Cornwell, P.J., Doebling, S.W., Prime, M.B.: Structural Health Monitoring Studies of the Alamosa Canyon and I-40 Bridges. Los Alamos National Laboratory Report LA-13635-MS (2000)

Chapter 5
Transmittance Anomalies for Model-Based Damage Detection with Finite Element-Generated Data and Deep Learning

Panagiotis Seventekidis and Dimitrios Giagopoulos

Abstract Numerically generated vibration responses may offer alternatives for the necessary data in structural health monitoring (SHM) applications that operate on damage detection and identification tasks. The main advantage of finite element (FE)-generated data is the substitution of costly and sometimes impossible experiments to acquire data for different healthy and damaged states. On the other hand, numerically generated data is strongly limited on the accuracy of models, meaning that it is as good as the accordance of the FE models with the possible real states. In the present work, a novel method is shown that focuses on generating SHM data with the use (FE) models by simulating the transmittance function (TF) anomalies that occur for a set of selected damage cases. TF function anomalies are approximated by simulating uncertainty on different model parameters and stiffness degradation on the corresponding areas of the selected damaged cases. The simulated dataset is used after for training of a Deep Learning (DL) Convolutional Neural Network (CNN) classifier. The presented methodology is tested on a lab scale CFRP truss structure for which different health scenarios are considered in a form of realistic damages on the CF truss members. The CNN classifier trained by numerically generated responses is finally validated on the generalization to the real experimental states. The simulated TF anomalies are found to be reliable for the test case as a training feature and can generalize on experimental states provided the feature space is enriched with uncertainty on the model parameters. The advantages and limitations are discussed.

Keywords Structural health monitoring · Damage identification · Finite element modeling · Deep learning

5.1 Introduction

Recent works on vibration SHM include model-based approaches [1–4] for the generation of necessary training datasets to be used in damaged detection or identification. FE models have been demonstrated to be able to substitute the physical measurements necessary to acquire the data for a wide range of excitations and health states. The advantages are presented as no need for actual physical damages and time/cost-intensive experiments. The main limitation, however, remains as how much the simulated data can be accurate in order to be able to train reliable SHM classifiers or decision systems, especially for small magnitude damages and early stages of detection. The points of attention for model-based SHM methods would therefore create two main areas of focus: First, what would be the nature of the utilized SHM features that generalize better between numerical and experimental responses for early detection? Second, how can the physics models account for a wide range of uncertainties in different future health states that remain unknown?

In the present work a novel model-based method is presented that utilizes FE-calculated TF anomalies as the SHM features for small damage detection and identification. The simulated TF anomalies are after used to train a CNN classifier as a decision system for the prediction of the real health state. The method is validated on experimental states of a CFRP truss for identification of different damaged truss members. Damage is induced on the truss members in a form of a relatively realistic and small magnitude impact-like scenarios.

TF functions [5–7] have already been proved to be sensitive in damage detection tasks and respond in stiffness changes of structures while being able to remain unaffected from the excitation. The TF function anomalies are introduced as a novel feature in this work for SHM tasks, calculated as the difference between the healthy reference and the inspected state. The advantage may be provided in potentially improved accuracy/generalization of the model-generated data since the deviation

P. Seventekidis · D. Giagopoulos (✉)
Department of Mechanical Engineering, University of Western Macedonia, Kozani, Greece
e-mail: pseventekidis@uowm.gr; dgiagopoulos@uowm.gr

R. Madarshahian, F. Hemez (eds.), *Data Science in Engineering, Volume 9*, Conference Proceedings of the Society for Experimental Mechanics Series, https://doi.org/10.1007/978-3-031-04122-8_5

from the healthy state is utilized, and a strict model fit to each state may not be necessary. The studied damage cases on the other hand involve complex CFRP impact-like damage which usually requires multilevel material models for accurate representation. However, an approximation is employed where the damaged members are modeled with reduced stiffness along all their length combined with uncertainty simulation on various model parameters and damage magnitudes. A rich dataset is therefore formed by including a wide range of parameters that may affect the TF anomalies for each health state. The reliability of the employed damage approximation is tested with comparison between datasets of different uncertainty ranges.

5.2 Background

The methodology followed in this work is presented in the following steps:

1. Construction of the FE model of the candidate structure and update of parameters on the healthy state
2. Simulation of the dynamic response of the structure for different damage scenarios with model parameter uncertainties. Damaged members are modeled with degraded stiffness. Extraction of vibration responses and calculation of transmittance deviations from the reference healthy state
3. Application of suitable DL classifiers trained with simulated transmittance deviations
4. Validation of the trained DL classifiers on experimental states. Assessment of their numerical to physical data generalization capability

The key components and theory are further discussed in this this section. First, for the model updating the CMA-ES [8], optimization algorithm is used to perform the FE fitting on reference experimental responses by minimizing the residual J:

$$J\left(\underline{\theta}\right) = \frac{1}{N}\sum_{i=1}^{N}\frac{\sum_{j=1}^{M}\left(\widehat{y}\left(\underline{\theta}\right)_{ij} - y_{ij}\right)^2}{\sum_{j=1}^{M}\left(y_{ij}\right)^2} \tag{5.1}$$

where $\widehat{y}\left(\underline{\theta}\right)_{ij}$ is the analytical acceleration time history computed from the numerical model and y_{ij} the experimental signal. The subscript i corresponds to the sensor and direction location out of a N total and j to the time step instant out of a total M. Second, the FE model is used for labeled vibration data generation of different health state scenarios including model parameters uncertainty following random sampling. The concept has been discussed in [1, 9] where each load case is solved after sampling on the parameters of the equation of motion in (5.2):

$$\mathbf{M}\left(\rho\right)\mathbf{A} + \mathbf{C}\left(\mathbf{K}, \mathbf{M}, \alpha, \beta\right)\mathbf{V} + \mathbf{K}(E)\mathbf{U} = \mathbf{F} \tag{5.2}$$

where **A**, **V**, and **U** are the global acceleration, velocity, and displacement vectors, respectively. **M**, **C**, and **K** represent the global mass, damping, and stiffness matrices of the structure that depend on the model physical parameters of density ρ, damping α, and β and elasticity E. For simulation of the damaged members, it is additionally assumed that the compromised members may be simulated with reduced stiffness E^d along their length. The reduced stiffness may be calculated as a percentage r of the original healthy stiffness as $E^d = rE^{\text{healthy}}$. A uniform reduction is applied to all material constants of the CFRP that form the vector as $E = [E_1\ E_2\ G_{12}\ G_{1Z}\ G_{2Z}]$ with the moduli of elasticity E_1 and E_2 as well the in-plane and transverse shear moduli G_{12}, G_{1Z}, and G_{2Z}. Finally, the constant r represents essentially the damage magnitude and can be regarded as one of the parameters subjected to uncertainty. After each simulation, the healthy and damaged states accelerations are used to compute the TFs between sensor pairs y_1 and y_2, defined as:

$$\text{TF}_{y_1y_2}\left(\omega\right) = \frac{S_{y_1y_2}\left(\omega\right)}{S_{y_2y_2}\left(\omega\right)} = \frac{x_{y_1}\left(\omega\right)\ x^*_{y_2}\left(\omega\right)}{x_{y_1}\left(\omega\right)\ x^*_{y1}\left(\omega\right)} \tag{5.3}$$

where the $S_{y_1y_2}$ is the cross-spectral density and $S_{y_2y_2}$ the auto-spectral density of the signals y_1 and y_2, while x and x^* represent the respective Fourier and conjugate Fourier transform of the corresponding signals. The transmittance anomaly then, or transmittance deviation, is defined for each damage case as: DTF $= |TF_{\text{healthy}} - TF_{\text{damage}}|$ where T_{healthy} and T_{test} represent the corresponding TFs of each structural health state. Algorithm 1 which is the complete data generation scheme may finally be presented to create a labeled training set of n cases by performing random sampling on the physical parameters of the model ρ, α, β, E, and r.

Algorithm 1: Numerical model data generation algorithm

Input: Number of load cases n and statistical bounds for each quantity ρ, α, β, E
Output: n number of labeled DTFs and health labels

1. **for** $i = 1 : n$ **do**
2. *sample* $E \to \mathbf{K} = \mathbf{K}(E)$
3. *sample* $\rho \to \mathbf{M} = \mathbf{M}(\rho)$
4. *sample* $\alpha, \beta \to \mathbf{C} = \mathbf{C}(\mathbf{K}, \mathbf{M}, a_i, \beta_i)$
5. *solve* $\mathbf{MA} + \mathbf{CU} + \mathbf{KU} = \mathbf{F}$
6. *calculate* $T_{healthy}$ *from* $\mathbf{A}$
7. *define Health status* $\to \mathbf{Y}_n$
8. *sample* $r \to$ *recalculate* E
9. *Solve* $\mathbf{MA} + \mathbf{CU} + \mathbf{KU} = \mathbf{F}$
10. *calculate* T_{damage} *from* $\mathbf{A}$
11. *return* $DTF_n = |T_{healthy} - T_{damage}|$ *and* $\mathbf{Y}_n$
12. **end**

The form of the training set containing the DTFs and health status labels $\mathbf{Y}_n$ is finally given in (5.4).

$$Train\_set = \{(DTF_1, \mathbf{Y}_1), (DTF_2, \mathbf{Y}_2), \ldots\ldots (DTF_n, \mathbf{Y}_n)\} \tag{5.4}$$

A CNN is trained then by the extracted training sets. During training or learning on the dataset, a CNN learns the appropriate filters that better reveal the damage-sensitive characteristics and patterns among sensors contained in the signals. For 1D signals as in the present work, a convolution filter $\mathbf{w}_k$ of a neuron k on an input signal $\mathbf{in}$ is defined as $\mathbf{f}_k$ according to (5.5) producing an output vector or filtered signal $(\mathbf{w}_k * \mathbf{in})[t]$ of length t.

$$\mathbf{f}_k = \mathbf{w}_k * \mathbf{in} \quad \text{with } (\mathbf{w}_k * \mathbf{in})[t] = \sum_j \mathbf{in}(j)\mathbf{w}_k(t - j + 1) \text{ (j takes all valid values)} \tag{5.5}$$

The parameters are learned in a way to minimize the difference between network output labels σ_i to true labels Y_i (for i number of classes) by minimizing the loss between predicted and actual labels using the categorical cross entropy of (5.6). Details can be found in [1, 10–12].

$$\mathrm{CE} = -\sum_{i=1} Y_i \ln(\sigma_i) \tag{5.6}$$

The trained CNN is finally validated on experimental inputs in order to predict the corresponding health state. A new metric is additionally introduced to assess the numerical to experimental generalization capabilities of the methodology. The mismatch between correct experimental and correct numerical health states inputs prediction by the trained CNN is given in (5.7). Low mismatch values indicate good generalization.

$$\text{mismatch} = \frac{\left|\text{Correct}_{\text{Experimental}} - \text{Correct}_{\text{Numerical}}\right|}{\text{Correct}_{\text{Numerical}}} \tag{5.7}$$

5.3 Analysis

The experimental setup of this work consists of the CFRP of Fig. 5.1 excited by random excitations from the electrodynamic shaker with an attached stringer rod. Accelerometers are placed on different locations in order to capture the vibration response.

The different health state scenarios are formed by substituting healthy truss members with damaged ones. The damage is enforced by 3-point bending to simulate impact like conditions. An indicative picture of the damage process is shown in Fig. 5.2 along with an example of a resulting damaged truss member.

The different damage cases for identification studied in this work are shown in Fig. 5.3. Each damage case from D1 to D6 is formed by substituting the corresponding healthy CFRP tube.

The magnitude of damage presents small changes in stiffness creating therefore a challenging early detection SHM problem. In Fig. 5.4, the calculated experimental TFs for different damaged cases are shown in comparison to the healthy state TF.

The FE model of structure along with the update procedure for the material parameters has been presented in [1]. Data is generated with the FE model by following Algorithm 1 using material parameters uncertainty a +/− 15% uniform random deviation from the updated values and stiffness reduction constant in a range of $r = 0.6 - 0.9$. The resulting simulated DTFs form a rich feature space based on the concept of parameters sampling. An indicative example for the Damage 1 case is shown in Fig. 5.5 where the experimental DTF is plotted along with the 5 randomly reelected simulated DTFs calculated by the FE model.

With the training set obtained by the FE simulations, a CNN is trained to perform damage identifaction hierarchically, meaning that different stages of damage separation are employed as discussed in [1]. In total, three stages are formulated in the studied problem with progressive refinement as shown in Fig. 5.6.

The typical architecture of the CNNs used is shown in Fig. 5.7. In total, 5 CNNs are used for the complete problem as impled by Fig. 5.6, namely, 1 CNN for Stage I, 2 CNNs for Stage II, and 2 as well for Stage III.

The numerically data-trained CNNs are finally validated on experimental inputs for prediction of the experimental state. Results are shown for Stage I in Fig. 5.8.

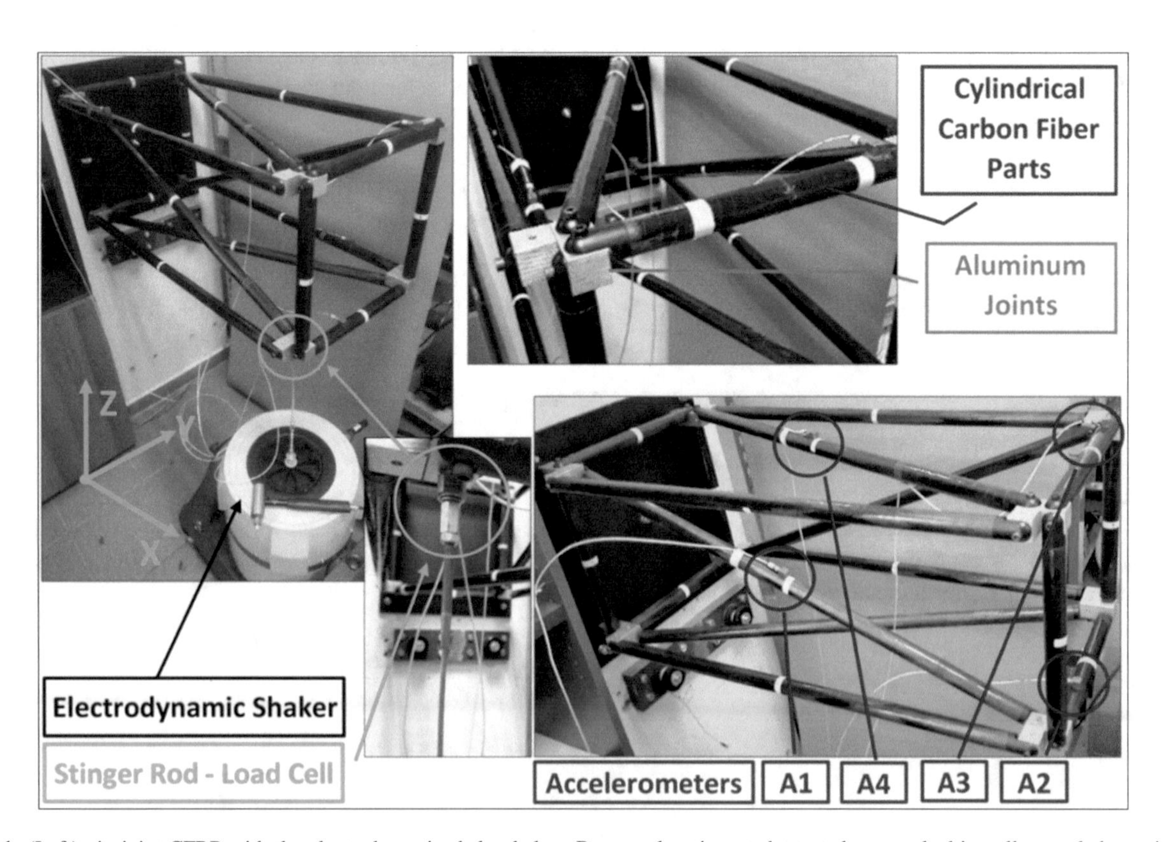

Fig. 5.1 (Left) pin-joint CFRP with the electrodynamic shaker below. Damage locations to be tested are marked in yellow and shown in more detail for the case of CFRP tube damages with the accelerometers locations (right)

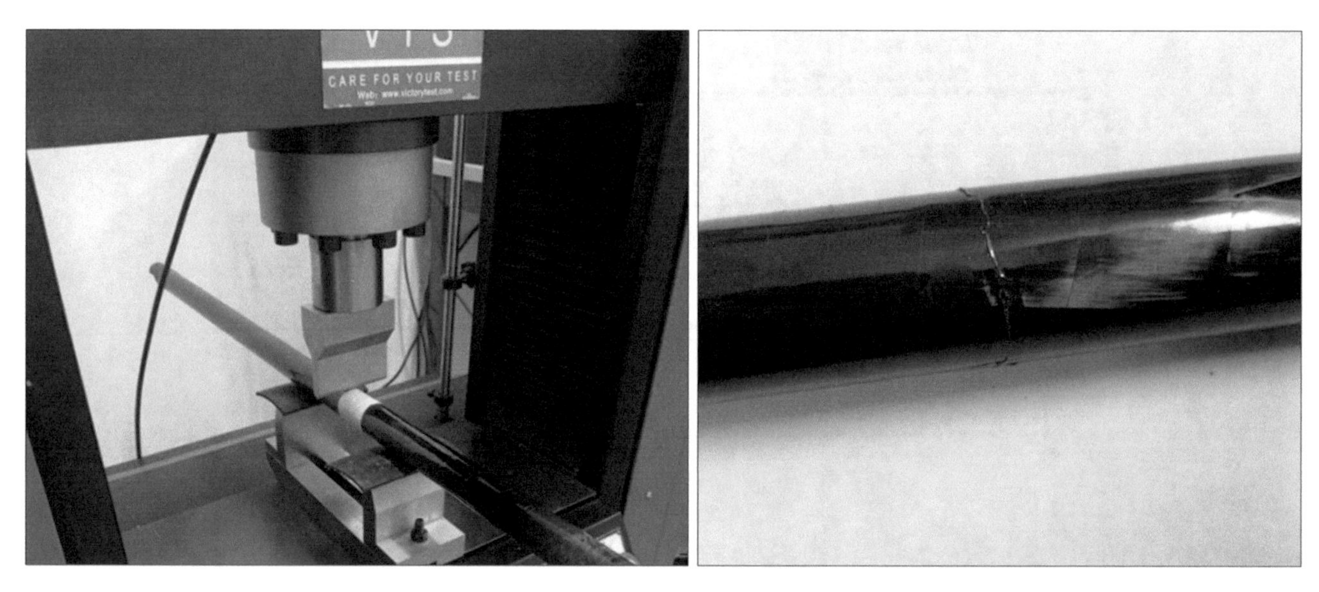

Fig. 5.2 Indicative picture of a 3-point bending damage application (left) and resulting truss member (right)

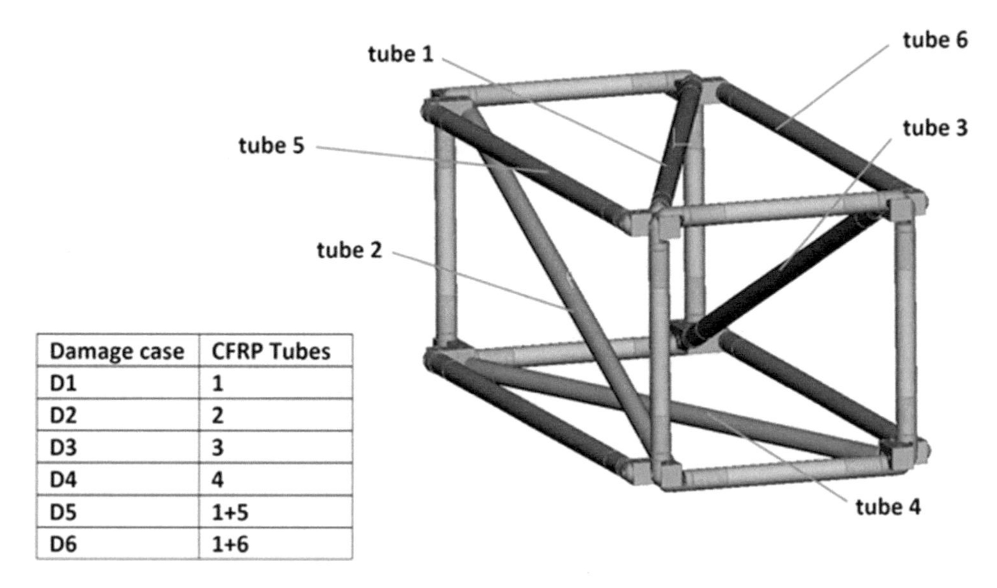

Damage case	CFRP Tubes
D1	1
D2	2
D3	3
D4	4
D5	1+5
D6	1+6

Fig. 5.3 Truss members naming convention used in damage cases considered

A number of 20 predictions on each state are indicatively shown for each D1–D6 states. The network prediction scores come in values between 0 and 1, with 0.5 being the class threshold. Blue is used for the score of the first class that the network predicts on the input and red for the second class. The sum of the blue and red column is always equal to 1. According to the damage identification problem, experimental states were classified correctly. D1, D2, and D5 were separated successfully from the D3, D4, and D6 cases suggesting that the features learned from the simulated data validate the experimental state too.

The advantage of using such features as DTFs instead of TFs or acceleration responses may be readily shown if the fit between the updated FE and experimental setup is compared, suggesting that no strict accordance was necessary to simulate correctly the TF deviations even for the small damages considered. In Fig. 5.9 the comparison between experimental and FE update TFs is shown.

On the other hand, the reader may notice that the methodology depends on random sampling of parameters to produce a rich feature space. The generalization capability of the numerically trained CNNs may be therefore limited on the quality of the fed-in data. To test this hypothesis, the data generation process is repeated using this time a $+/-$ 3% uniform random deviation from the updated values instead of $+/-$ 15%. The prediction scores of the corresponding CNNs on the same experimental inputs are given in Fig. 5.10. The performance this time, however, suffers with less confident and accurate

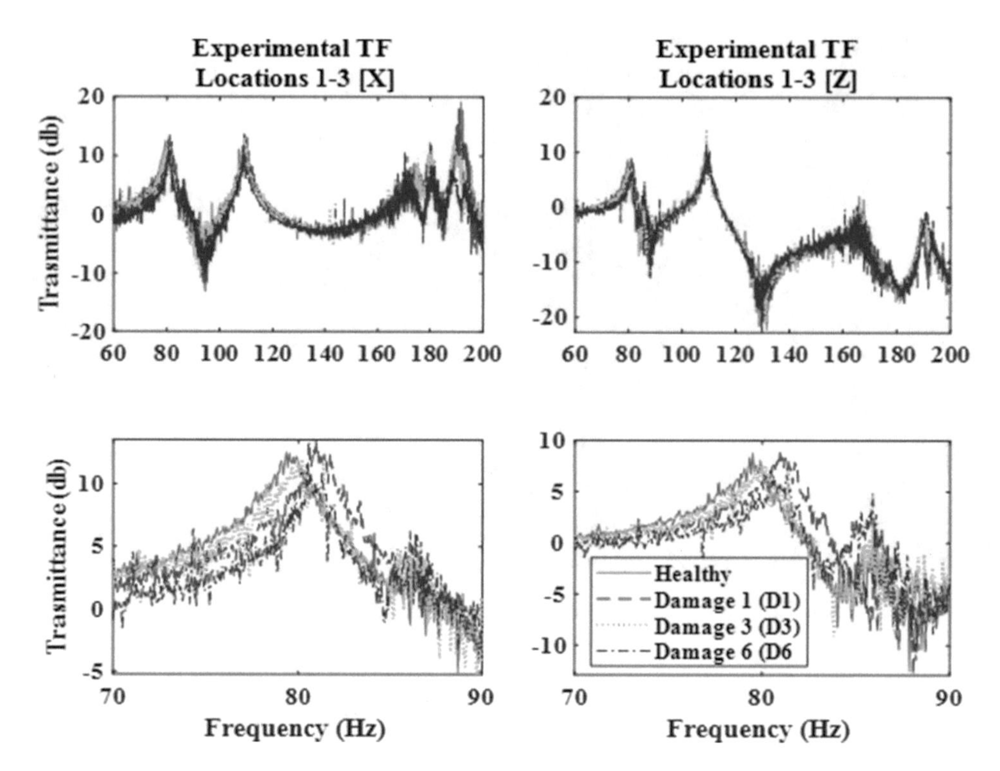

Fig. 5.4 Experimental TFs between sensor locations 1–3 for a range of 60–200 Hz (above) and 70–90 Hz (below) for the healthy and damage 1, 2, and 3 cases

predictions on the experimental states and confirms the dependence of generalization on the uncertainty and necessary enrichment of numerical data. Table 5.1 shows a comparison between the generalizations to the experimental states for the two different datasets used in this work, using as index the mismatch defined in (5.7). For the numerical inputs' correct predictions, the scores of validation during training of the CNNs on the corresponding FE datasets were used. For the experimental inputs' correct predictions, the scores were drawn with the same procedure as in Figs. 5.8 and 5.9.

5.4 Conclusion

In this work, a model-based method using FE-generated vibration data to calculate DTFs datasets for SHM training was tested on a CFRP pin-joint truss with impact-like experimental damage. The DTF data was used to train CNN classifiers which showed to be able to generalize correctly on the real structure even though no strict fit was achieved between the healthy numerical FE model and the experimental state. The damage was approximated by uniform stiffness reduction on the corresponding truss member which showed to be adequate, even though simple, provided however that the model parameter uncertainties were considered in a wide range.

Acknowledgments We acknowledge support of this work by the project "Development of New Innovative Low-Carbon Energy Technologies to Enhance Excellence in the Region of Western Macedonia" (MIS 5047197) which is implemented under the Action "Reinforcement of the Research and Innovation Infrastructure", funded by the Operational Program "Competitiveness, Entrepreneurship and Innovation" (NSRF 2014-2020) and co-financed by Greece and the European Union (European Regional Development Fund).

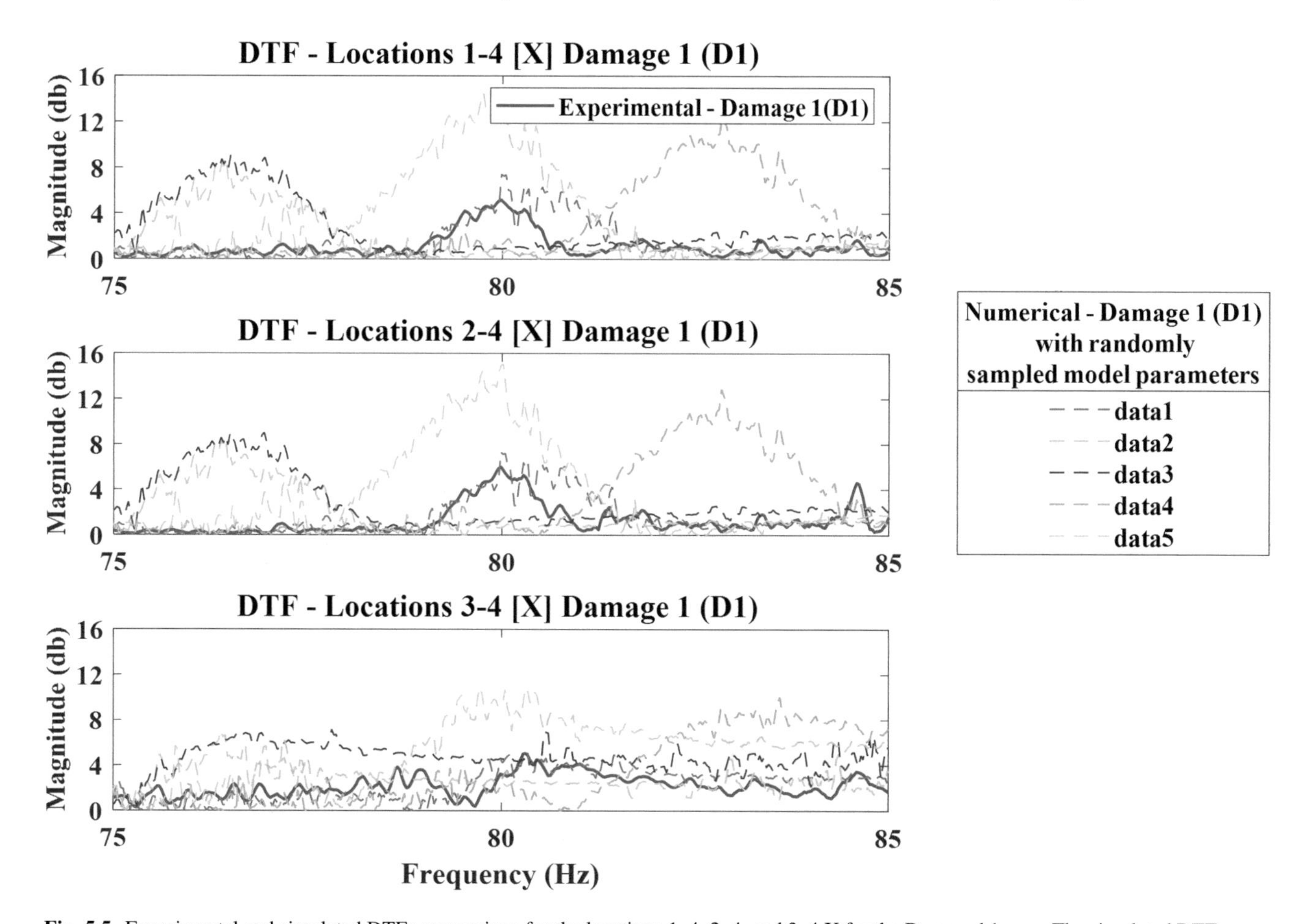

Fig. 5.5 Experimental and simulated DTFs comparison for the locations 1–4, 2–4, and 3–4 X for the Damaged 1 case. The simulated DTFs were generated with randomly sampled material parameters

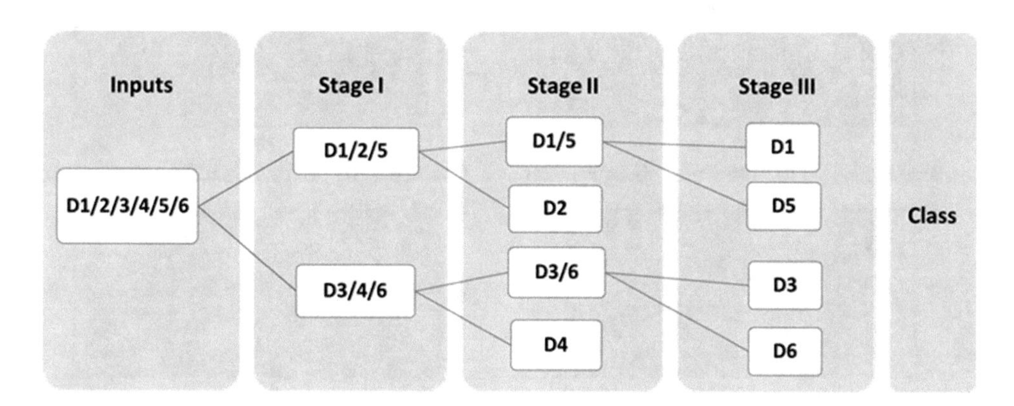

Fig. 5.6 Hierarchical separation of damages

Table 5.1 Generalization from numerical to experimental data in terms of mismatch between correct classification scores for numerical and experimental inputs

	Highest mismatch in correct predictions between numerical and experimental inputs for Stage I	
Dataset used	Mismatch %	Damage case
$r =$ **0.6–0.9** and **15**% material uncertainty	3	D2
$r =$ **0.6–0.9** and **3**% material uncertainty	31	D1

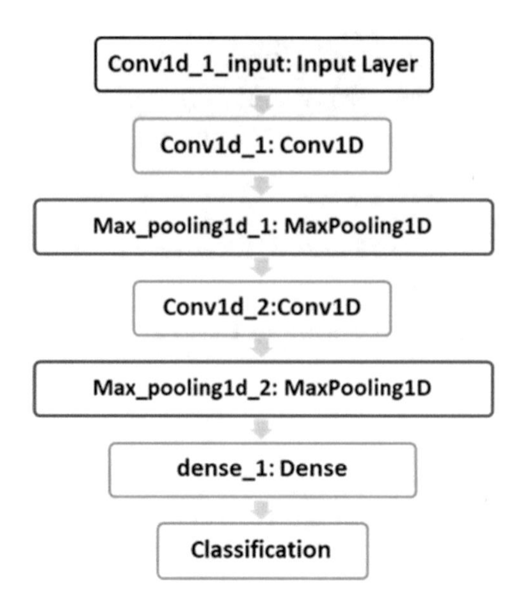

Fig. 5.7 CNN classifier architecture used

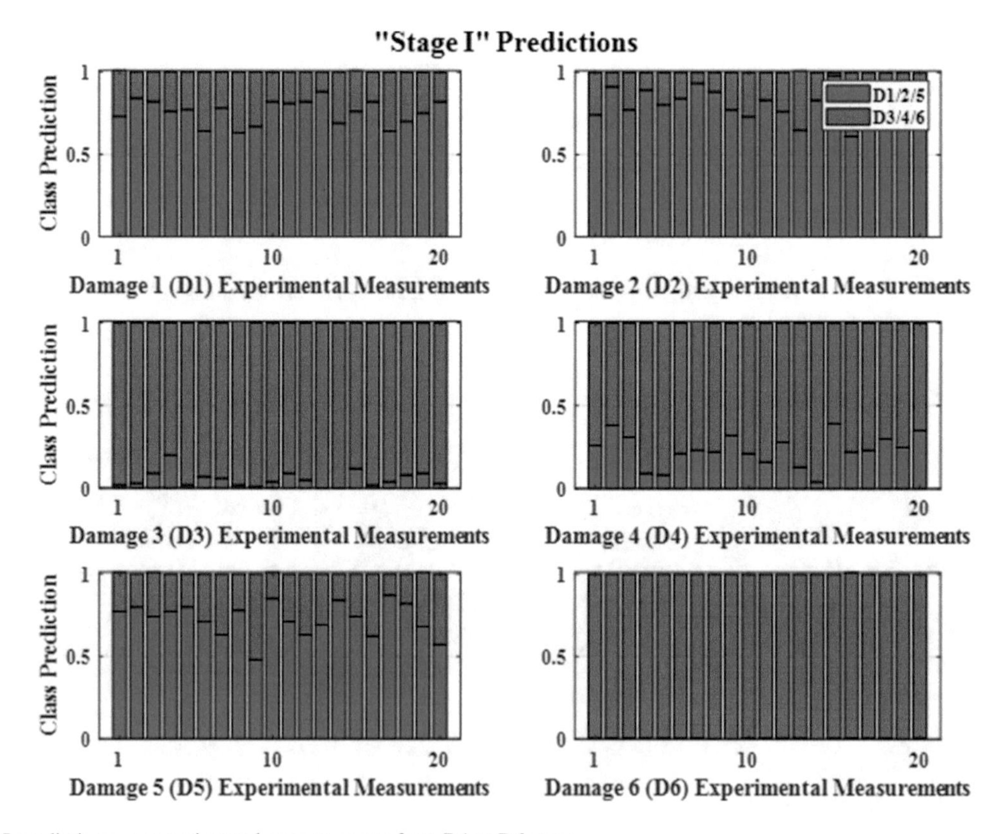

Fig. 5.8 Stage I predictions on experimental measurements from D1 to D6 states

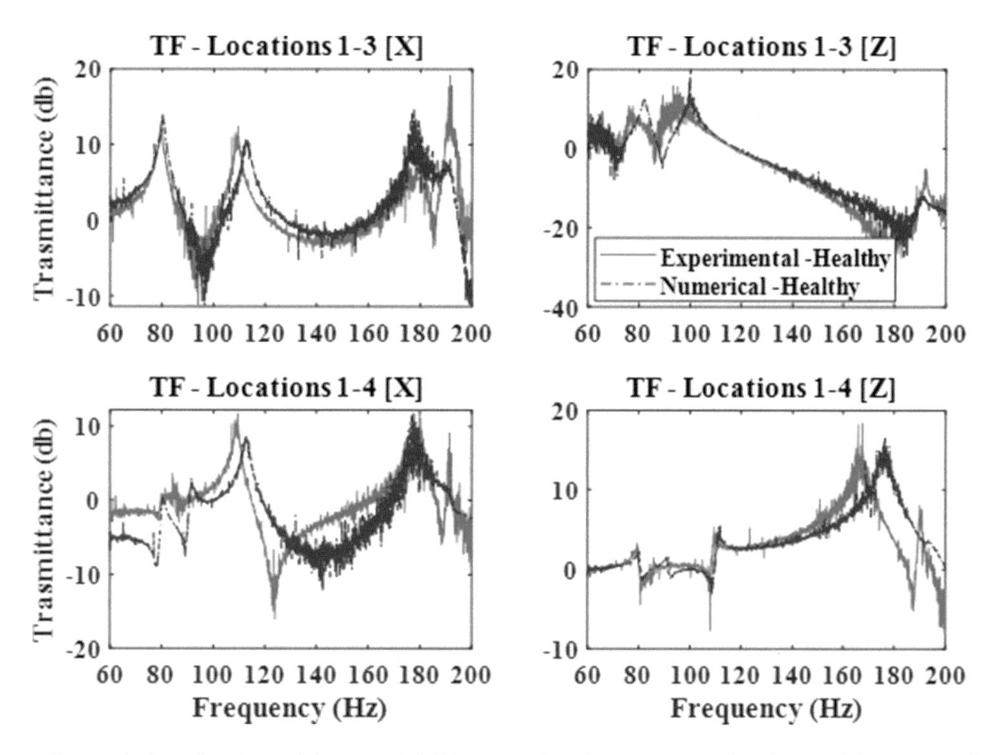

Fig. 5.9 Experimental-numerical (updated material properties) TFs comparison between sensor locations 1–4 for a range of 70–90 Hz for the healthy cases

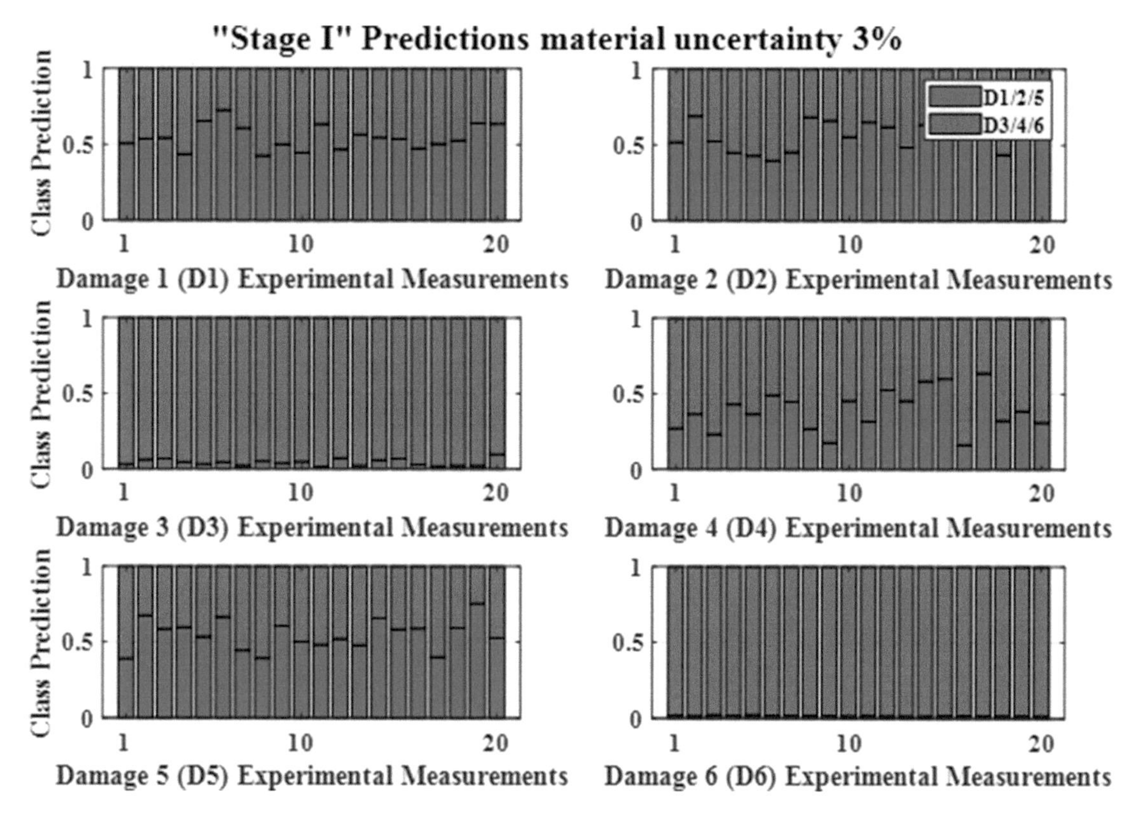

Fig. 5.10 Stage I predictions on experimental measurements from D1 to D6 states using material uncertainty simulation of 3%

References

1. Seventekidis, P., Giagopoulos, D.: A combined finite element and hierarchical Deep learning approach for structural health monitoring: test on a pin-joint composite truss structure. Mech. Syst. Signal Process. **157**, 107735 (2021)
2. Mousavi, Z., Varahram, S., Ettefagh, M.M., Sadeghi, M.H., Razavi, S.N.: Deep neural networks–based damage detection using vibration signals of finite element model and real intact state: an evaluation via a lab-scale offshore jacket structure. Struct. Health Monit. **20**(1), 379–405 (2021)
3. Barraza, J.F., Droguett, E.L., Naranjo, V.M., Martins, M.R.: Capsule Neural Networks for structural damage localization and quantification using transmissibility data. Appl. Soft Comput. J. **97**, 106732 (2020)
4. Padil, K.H., Bakhary, N., Abdulkareem, M., Li, J., Hao, H.: Non-probabilistic method to consider uncertainties in frequency response function for vibration-based damage detection using Artificial Neural Network. J. Sound Vib. **467**, 115069 (2020)
5. Poulimenos, A.G., Sakellariou, J.S.: A transmittance-based methodology for damage detection under uncertainty: an application to a set of composite beams with manufacturing variability subject to impact damage and varying operating conditions. Struct. Health Monit. **18**(1), 318–333 (2019)
6. Zhang, H., Schultz, M.J., Ferguson, F.: Structural health monitoring using transmittance functions. Mech. Syst. Signal Process. **13**(5), 765–787 (1999)
7. Caccese, V., Mewer, R., Vel, S.S.: Detection of bolt load loss in hybrid composite/metal bolted connections. Eng. Struct. **26**, 895–906 (2004)
8. Giagopoulos, D., Arailopoulos, A.: Computational framework for model updating of large scale linear and nonlinear finite element models using state of the art evolution strategy. Comput. Struct. **192**, 210–232 (2017)
9. Seventekidis, P., Giagopoulos, D., Arailopoulos, A., Markogiannaki, O.: Structural health monitoring using deep learning with optimal finite element model generated data. Mech. Syst. Signal Process. **145**, 106972 (2020)
10. Hagan, M.T., Demuth, H.B., Beale, M.H., De Jesus, O.: Neural Network Design. PWS Publishing Company, Boston (2014)
11. Aggarwal, C.C.: Neural Networks and Deep Learning. A Textbook 2018. Springer. 978-3-319-94463-0
12. Abdeljaber, O., Avci, O., Kiranyaz, S., Gabbouj, M., Inman, D.J.: Real-time vibration-based structural damage detection using one-dimensional convolutional neural networks. J. Sound Vib. **388**, 154–170 (2017)

Chapter 6
Machine Learning-Based Condition Monitoring with Multibody Dynamics Models for Gear Transmission Faults

Josef Koutsoupakis, Panagiotis Seventekidis, and Dimitrios Giagopoulos

Abstract In this work, a condition monitoring (CM) data generation scheme is presented for a two-stage gear transmission system. Simulated vibration response data is used to train a convolutional neural network (CNN) which performs damage identification on two experimental damaged states. The multibody dynamics (MBD) model of a two-stage helical gear transmission is first developed and used to model the healthy and the damaged state of the problem. Data is afterwards generated through an uncertainty simulation repetitive load case algorithm and is found to be able to train an accurate CM-CNN classifier which validates experimentally measured structure states. The proposed method aims to bridge the gap between digital and experimental data for CM or structural health monitoring (SHM). The framework may find applications in various cases were experimental data for training CM or SHM classifiers is sparse or the cost of experiments is high.

Keywords Deep learning · Neural network · Condition monitoring · Multibody dynamics · Gear transmission

6.1 Introduction

Modern needs call for adoption of advanced damage detection and identification schemes for structural and mechanical systems. Condition monitoring (CM) and structural health monitoring (SHM) have been referred widely as terms that comprise methodologies of damage detection and identification in an implicit way, meaning without physically observing or locating the damage and anomalies. The most common CM and SHM methodologies include detailed vibration measurements that are used to detect changes in a mechanical system's behavior and identify damage. Cases in literature include damage detection in various structural [1–3] and mechanical systems like gear transmissions [4–6]. More recently, methodologies for SHM have been combined with machine learning (ML) methods where a mathematical model is built and trained by structural response data [7–9] and learns to recognize potential anomalies or damage in measured signals. The measured response has however to be processed initially by the user, in order to extract the appropriate damage-sensitive features of the signal and afterwards feed them to the classifier.

A class of ML methods known as artificial neural networks (ANN) when applied in a deep or deep learning (DL) form can learn the appropriate signal features and damage thresholds automatically. Therefore, great advantage of limited user preprocessing in arbitrary damage cases is implemented with DL. A category of DL classifiers known as convolutional neural networks (CNN) has already been proposed for SHM or CM applications trained directly by experimentally measured responses [10, 11]. More recently, CNNs have also been proposed to be trained by numerically generated finite element (FE) responses and be used in turn for SHM on real structures [12]. The great advantage of CNNs to be able to generalize on physical structures while trained numerically reveals a potential of acquiring data for structures and mechanical systems where it was not possible or experimenting with inducing actual damage is prohibited. Examples of such cases can be expensive or large structures such as bridges where deliberately damaging [2] the structure to get data is not advised or the experiment cost is high. In addition, the approach may also be of use for CM where large datasets in various ambient and working conditions of moving mechanical components are necessary for acquisition of the appropriate data. Thus, with a digitally trained DL classifier, in cases where numerical simulations to acquire data are feasible, both the user prepressing/input and the problem of possible lack of experimental training data are tackled. The numerically generated data may also label the damage types to aid decisions for quick maintenance or service life planning.

J. Koutsoupakis · P. Seventekidis · D. Giagopoulos (✉)
Department of Mechanical Engineering, University of Western Macedonia, Kozani, Greece
e-mail: pseventekidis@uowm.gr; dgiagopoulos@uowm.gr

R. Madarshahian, F. Hemez (eds.), *Data Science in Engineering, Volume 9*, Conference Proceedings of the Society for Experimental Mechanics Series, https://doi.org/10.1007/978-3-031-04122-8_6

For the digital SHM or CM data generation, numerical models have of course to be as accurate as possible. Updated or optimal FE/MBD models are fit to experimentally measured data [13, 14] that could be time or frequency response of the system reproducing afterwards the given experimental state with good accuracy. The user may afterwards consider the updated state as the healthy one and simulate [12] the required damage cases with the requirement that they are adequately defined, e.g., fatigue cracks, loose bolts, damaged gear teeth. Therefore, after an optimal model of a structure is built by an initial experimental cost on the healthy structure state, literally no limitation arises for CM or SHM training data generation. Model uncertainties could also be simulated additionally in order to teach the classifier to generalize better [12]. A trained CNN can operate after with a permanently installed network of sensors which often records input-output or output-only vibration measurements during operation of a structure, both before and after a damage event.

The aim of this work is to present an uncertainty simulating FE/MBD data-trained CM-CNN classifier for a two-stage gear transmission system in order to extend the promising attempt on a small-scale linear structure [12]. The proposed data generation framework will be validated for CM of gear pairs, and the potential to be used as a future tool for damage identification will be tested. The proposed framework may find application both in lab and real-world tasks.

The presentation follows with the Background section where the theoretical background of the process is described. The Analysis section presents the experimental setup and the FE/MBD model and a description of the imposed damages for the experimental and the numerical system. The predictions of the trained CNN on experimental measurements for healthy and damaged gear states are presented in order to validate the CM-CNN method experimentally. Finally, in the Conclusions section, the results are discussed.

6.2 Background

In this section, the steps followed in the present CM-CNN methodology are summarized. The process is shown described in the flowchart of Fig. 6.1.

For a rigid-flexible MBD system, the equations of motion can be written in the following general form [15]:

$$\begin{cases} M\ddot{q} + \Phi_q^T \lambda + F_q = Q(q) \\ \Phi(q, t) = 0 \end{cases} \tag{6.1}$$

where M is the system mass matrix; $\Phi(q, t)$ is the vector that contains the system constraint equations corresponding to the ideal joints; t represents the time; Φ_q is the derivative matrix of constraint equations with respect to the system generalized coordinates q; λ is the Lagrange multipliers associated with the constraints; $Q(q)$ is the system external generalized forces, e.g., the gravity force, other external spring force, and damping force; and $F(q)$ denotes the system elastic force vector. Contacts between bodies can be defined using the Hertzian contact force model along with the Coulomb friction model to calculate the total contact force as:

$$F_C = F_n\left(K, C, g, e\right) + F_r\left(\mu_s, \mu_d\right) \tag{6.2}$$

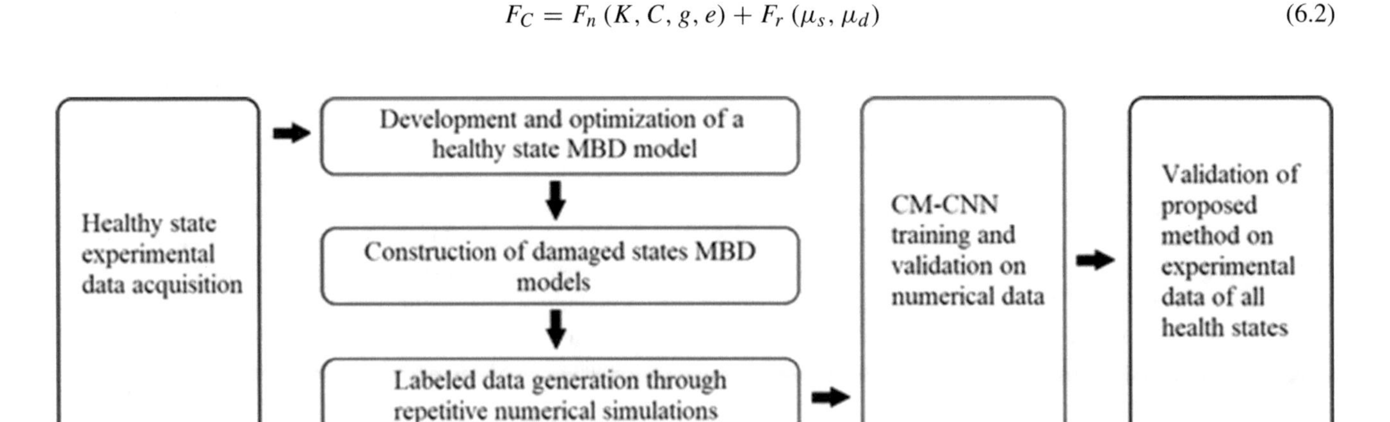

Fig. 6.1 Flowchart of proposed framework

where F_n and F_r are the normal contact force and the friction force, respectively. The terms K, C, g, e, μ_s, and μ_d denote the contact stiffness, damping, penetration depth, and static and dynamic friction coefficients, respectively.

Labeled datasets can be generated using an optimal MBD model [13], by running repetitive simulations [10] as shown in Algorithm 1. The contact parameters K, C, g, e, μ_s, μ_d, and external loads Q are randomly sampled for N cases [16], producing N-labeled CM datasets for the CNN training.

Algorithm 1

Input: N statistical bounds vectors for contact parameters K, C, g, e, μ_s, μ_d, and loads Q
Output: N-labeled acceleration time histories at prescribed nodes

1. *for i = 1:N*
2. *Define model's health status Y*
3. *Sample* K, C, g, e, μ_s, μ_d and Q
4. *Calculate contact force* F_c
5. *Solve equations of motion:*
$$\begin{cases} M\ddot{q} + \Phi_q^T \lambda + F_q = Q(q) \\ \Phi(q, t) = 0 \end{cases}$$
for $\ddot{q}$
6. *Return* $\ddot{q}$ *and Y*
7. end

The health label Y for each individual case is a single integer number that denotes the simulated health state and is related to the specific MBD model solved, being the healthy or some damaged instance. Finally, the numerical data is grouped to form the complete labeled dataset as shown in Eq. 6.3:

$$Train\_set = \left\{ \left(\ddot{q}_1, Y_1\right), \left(\ddot{q}_2, Y_2\right), \ldots \left(\ddot{q}_n, Y_n\right) \right\} \tag{6.3}$$

Using this dataset, a 1D CNN can be trained by applying a filter w_k on the 1D input a neuron k. The filtered output f_k of length t is calculated according to the convolution process:

$$f_k = w_k * in \text{ with } (w_k * in) = \sum_j in(j) w_k (t - j + 1) \ \text{(j takes all valid values)} \tag{6.4}$$

The CNN is trained by adjusting its parameters based on the minimization of the cross-entropy between the predicted and true labels of the data as:

$$\text{CE} = -\sum_{i=1}^{N} Y_i \ln\left(\hat{Y}_i\right) \tag{6.5}$$

where Y_i is the true label of the input data, $\hat{Y}_i$ is the network's prediction, and N is the number of input datasets.

6.3 Analysis

The experimental device Drivetrain Prognostics Simulation – DPS (Spectra Quest Inc.) is shown in Fig. 6.2. The setup components are described as follows: (1) electric motor, (2) single-stage planetary gear system with a 27:1 ratio, (3) two-stage gear system with a 2.5 maximum ratio per stage (6.25 max), (4) resistance-load gear boxes connected on the resistance-load inducing electric motor, (5) and (6) electric control for the setup. Focus in this work is given on the planetary and two-stage gearbox system.

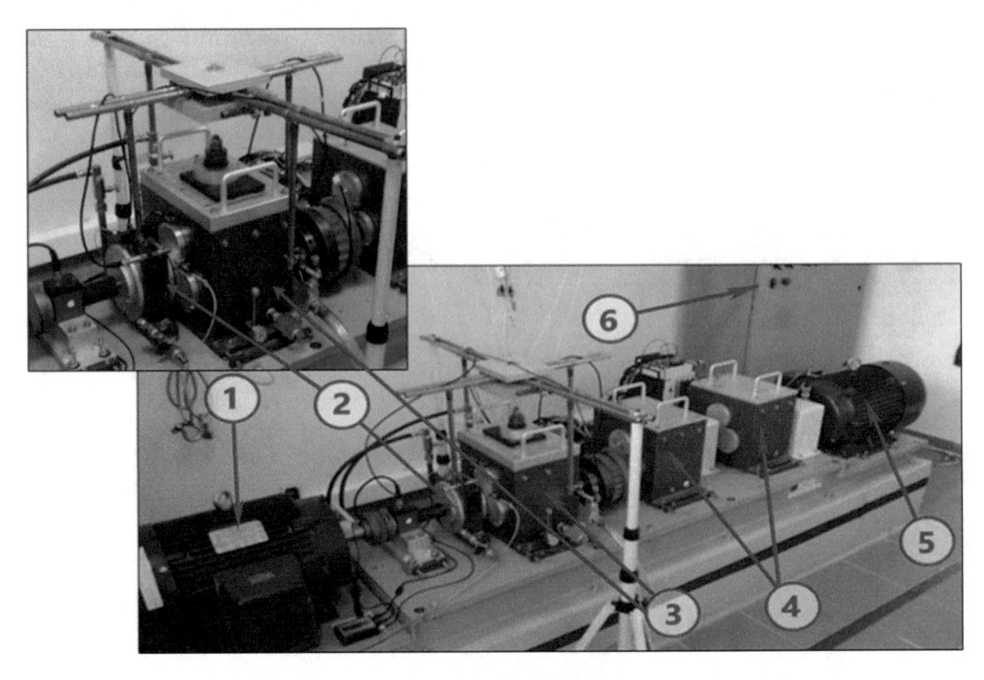

Fig. 6.2 Experimental device Drivetrain Prognostics Simulation – DPS (SpectraQuest Inc.)

Fig. 6.3 (Left) damaged helical gear with the missing tooth fault and (right) damaged helical gear with the surface fault

Three health states are examined here. The healthy state, where all system components are fully intact, will be referred as the Healthy. The two damaged states are created by inducing damage in a helical gear of the two-stage gearbox. The first damage case, a gear with a missing tooth, will be referred as the Damage 1 (D1) case. The second damage case, a gear with surface fault, will be referred as the Damage 2 (D2) case. Both cases are shown in Fig. 6.3.

Measurements of acceleration time histories are obtained using eight (8) accelerometers placed on specific positions of the gearbox as shown in Fig. 6.4. Two accelerometers are placed on the planetary gear system (A1, A2) and six more are placed on the two-stage gear system (A3–A8), placed as close to the bearings as possible. For all three health states, 100 seconds measurements are acquired at a rate of 10240 Hz, at 612, 911, and 1210 RPM. The setup operates at a constant speed and 10% load. The data collected for all cases is used to validate the CM-CNN trained by the numerical data, while the Healthy state data is also used for the optimization and updating of the multibody model.

For the numerical data that will be used to train the CM-CNN classifier, a MBD model of the two-stage gearbox is built as shown in Fig. 6.5. To bridge the gap between the experimental measurements and the numerical data, the model is optimized using the covariance matrix adaptation evolution strategy (CMA-ES), trying to minimize a cost function of the form:

Fig. 6.4 Accelerometer sensor positions on the two-stage gearbox

Fig. 6.5 The complete multibody and flexible parts for the two-stage gearbox of the benchmark problem with the single planetary gearbox

$$J\left(\underline{\theta}\right)=\sum_{i=1}^{M}\sqrt{\sum_{j=1}^{N}\left(\hat{y}_{ij}\left(\underline{\theta}\right)-y_{ij}\right)^{2}} \tag{6.6}$$

where $\hat{y}_{ij}\left(\underline{\theta}\right)$ and y_{ij} are the scaled power spectral densities (PSDs) of the numerical and experimental acceleration data, respectively, $\underline{\theta}$ denotes the vector containing the contact parameters K, C, g, e, μ_s, μ_d, and external forces Q. The subscripts i, j correspond to the accelerometer axes and frequency value, respectively.

The Damage 1 state model is constructed by altering the CAD geometry of the healthy gear to create the missing tooth case. For the Damage 2 state, the contact stiffness and friction coefficients are altered, to simulate the surface fault. The optimal contact parameters are displayed in Fig. 6.6, along the respective gear geometry for each case. Contacts 1 and 2 are defined between the gears of the two-stage gearbox, with Contact 2 being the one between the damaged gear and the outlet shat gear. The external load is the same for all cases, estimated at 68086 Nmm.

Figure 6.7 shows the model during a dynamic analysis. The colored contours display the flexible body deformations. Simulations are performed at a timestep value inverse to the sampling frequency. This way the acceleration histories at the specified accelerometer positions are extracted with the same sampling frequency as the experimental data.

The PSDs are calculated based on Welch's method [17] for each dataset and used to train the CM-CNN. Focus is placed on the first gear mesh frequency (GMF) of each inlet speed case, calculated as:

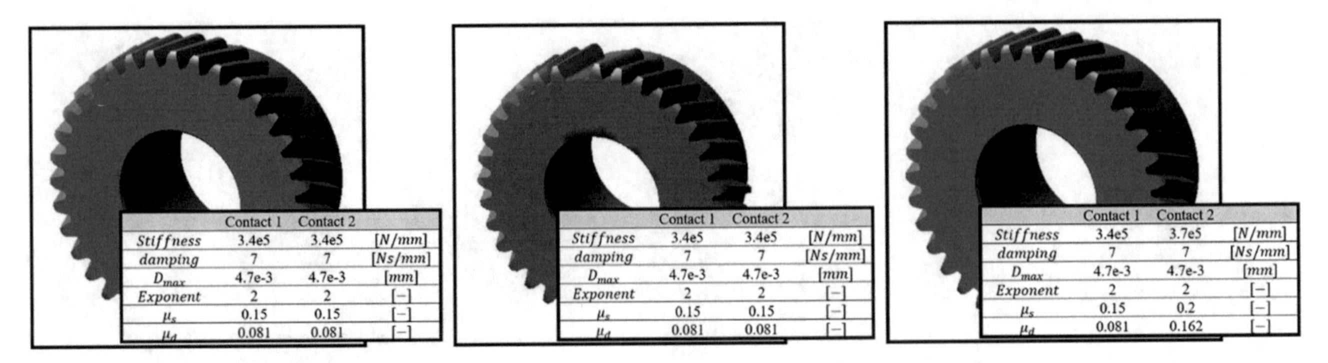

	Contact 1	Contact 2	
Stiffness	3.4e5	3.4e5	[N/mm]
damping	7	7	[Ns/mm]
D_{max}	4.7e-3	4.7e-3	[mm]
Exponent	2	2	[−]
μ_s	0.15	0.15	[−]
μ_d	0.081	0.081	[−]

	Contact 1	Contact 2	
Stiffness	3.4e5	3.4e5	[N/mm]
damping	7	7	[Ns/mm]
D_{max}	4.7e-3	4.7e-3	[mm]
Exponent	2	2	[−]
μ_s	0.15	0.15	[−]
μ_d	0.081	0.081	[−]

	Contact 1	Contact 2	
Stiffness	3.4e5	3.7e5	[N/mm]
damping	7	7	[Ns/mm]
D_{max}	4.7e-3	4.7e-3	[mm]
Exponent	2	2	[−]
μ_s	0.15	0.2	[−]
μ_d	0.081	0.162	[−]

Fig. 6.6 Contact parameters and gear geometry of (**a**) Healthy case, (**b**) Damage 1 case, and (**c**) Damage 2 case

Fig. 6.7 Flexible body deformations on the gearboxes (left) and on the two-stage gearbox's shafts (right)

$$\text{GMF} = Z_g f_g \tag{6.7}$$

where Z_g denotes the number of gear teeth and f_g the gear's rotational frequency. When damage is present in a system, anomalies appear on the GMFs. These changes in the spectrum include differences in amplitude as well as the appearance of sidebands around the GMFs [18–20]. A comparison of the experimental and optimized numerical frequency response (PSD) of the of the system is shown on Fig. 6.8, for frequency values around the first GMF in each RPM case and for all axes of A8.

The CM-CNN performance on the experimental data is asserted by its prediction accuracy on 100 samples for each health case. Figures 6.9, 6.10 and 6.11 present the certainty at which the network makes the predictions for 10 randomly selected samples of each health state and inlet speed case. In the same figures, the corresponding confusion matrix of each case is shown for all 300 experimental datasets. The prediction problem between these three health states is a multi-classification problem, with the threshold of prediction for damage presence at 33.3%. The network is shown to predict the health state of the experimental setup with large certainty and accuracy in each case, proving the robustness and reliability of the proposed method. As seen from the charts, the network has difficulty discerning the second damaged state from that of the healthy one as the rotational speed increases, while the Healthy and Damage 1 states are almost always accurately asserted. The highest overall accuracy is achieved at the low speed of 612 rpm at 98.7%. This indicates that predictions at lower rpm values provide more reliable predictions.

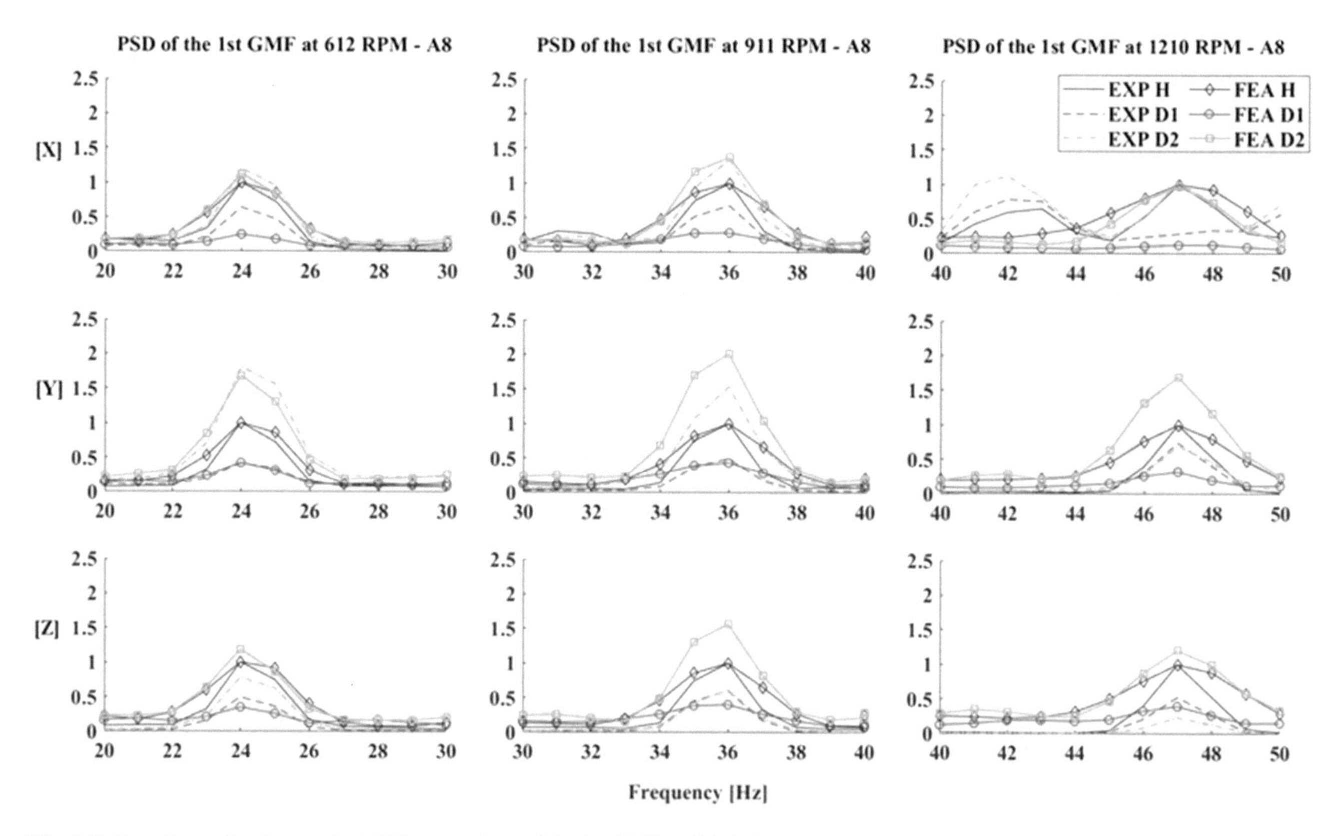

Fig. 6.8 Experimental and numerical PSD comparison of the 1st GMF at 612 (left), 911 (middle), and 1210 RPM (right)

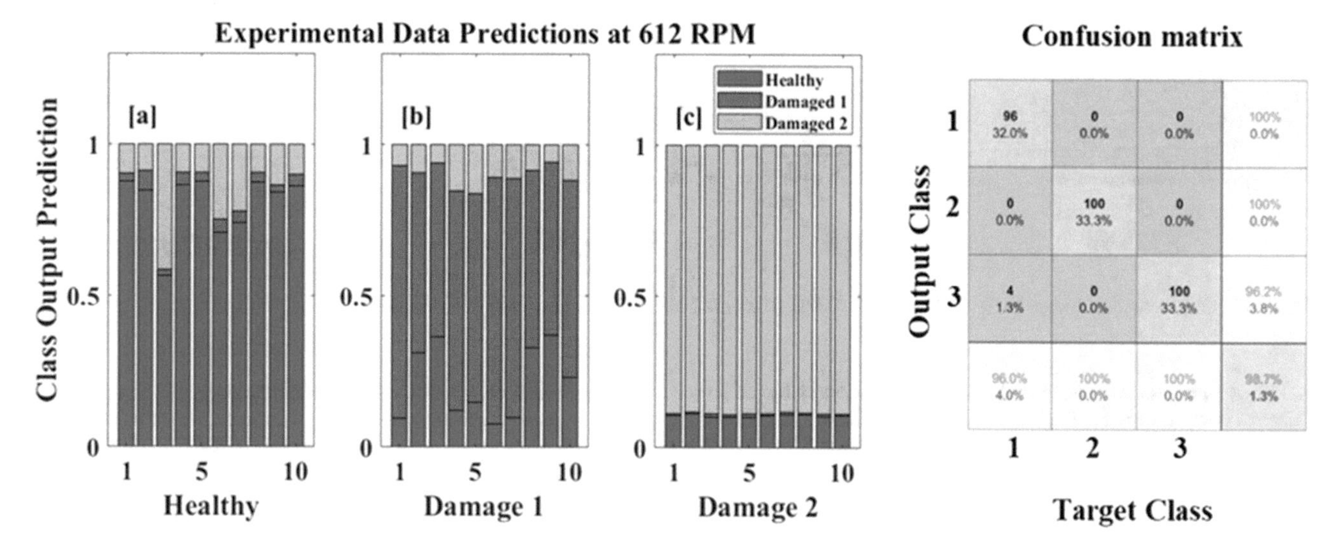

Fig. 6.9 CM-CNN prediction values (left) and confusion matrix (right) at 612 RPM

6.4 Conclusion

In this work, a data generation framework for CM of a two-stage gearbox through the corresponding FE/MBD model was presented. The data was then used to train a CNN classifier to perform damage identification on the real two-stage gearbox structure. The trained classifier proved to be able to separate the Healthy from the two Damaged states, the latter being a missing tooth and a surface fault on a helical gear. The framework will be further tested for different speed and load setups as well as different damages on the mechanical components.

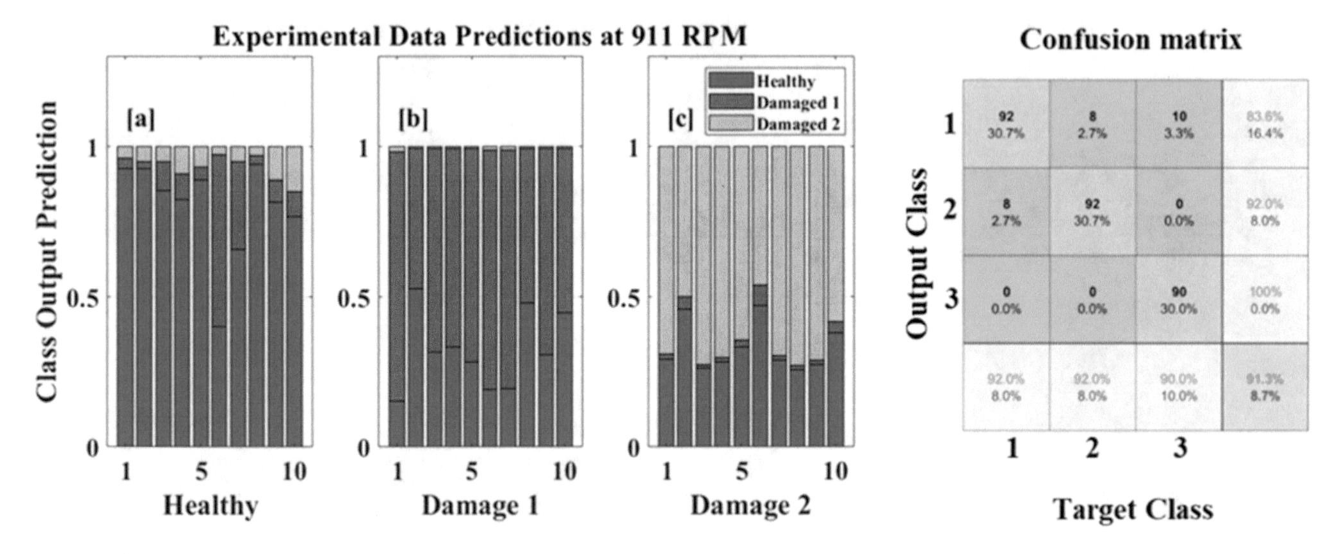

Fig. 6.10 CM-CNN prediction values (left) and confusion matrix (right) at 911 RPM

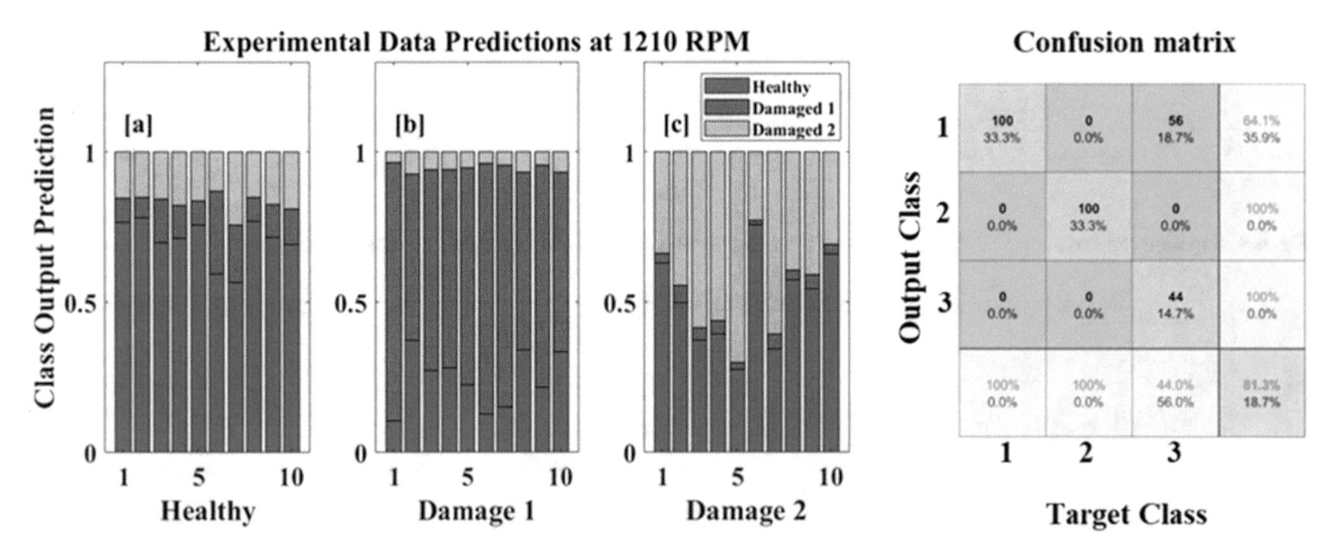

Fig. 6.11 CM prediction values (left) and confusion matrix (right) at 1210 RPM

Acknowledgments We acknowledge support of this work by the project "Development of New Innovative Low-Carbon Energy Technologies to Enhance Excellence in the Region of Western Macedonia" (MIS 5047197) which is implemented under the Action "Reinforcement of the Research and Innovation Infrastructure," funded by the Operational Program "Competitiveness, Entrepreneurship and Innovation" (NSRF 2014-2020) and co-financed by Greece and the European Union (European Regional Development Fund).

References

1. Wickramasinghe, W.R., Thambiratnam, D.P., Chan, T.H.T., Nguyen, T.: Vibration characteristics and damage detection in a suspension bridge. J. Sound Vib. **375**, 254–274 (2016)
2. Döhler, M., Hille, F., Mevel, L., Rücker, W.: Structural health monitoring with statistical methods during progressive damage test of S101 Bridge. Eng. Struct. **69**, 183–193 (2014)
3. Tcherniak, D., Mølgaard, L.L.: Active vibration-based structural health monitoring system for wind turbine blade: demonstration on an operating Vestas V27 wind turbine. Struct. Health Monit. **16**(5), 536–550 (2017)
4. Elasha, F., Greaves, M., Mba, D.: Planetary bearing defect detection in a commercial helicopter main gearbox with vibration and acoustic emission. Struct. Health Monit. **17**(5), 1192–1212 (2018)
5. Hu, W., Chang, H., Gu, X.: A novel fault diagnosis technique for wind turbine gearbox. Appl. Soft Comput. J. **82**, 105556 (2019)

6. Fuentes, R., Dwyer-Joyce, R.S., Marshall, M.B., Wheals, J., Cross, E.J.: Detection of sub-surface damage in wind turbine bearings using acoustic emissions and probabilistic modelling. Renew. Energ. **147**, 776–797 (2020)
7. Fassois, S.D., Kopsaftopoulos, F.P.: Statistical time series methods for vibration based structural health monitoring. New Trends Struct. Health Monit. **542**, 209–264 (2013)
8. Bornn, L., Farrar, C.R., Park, G., Farinholt, K.: Structural health monitoring with autoregressive support vector machines. J. Vib. Acoust. Trans. ASME. **131**(2), 0210041–0210049 (2009)
9. Mechbal, N., Uribe, J.S., Rébillat, M.: A probabilistic multi-class classifier for structural health monitoring. Mech. Syst. Signal Process. **60**, 106–123 (2015)
10. Abdeljaber, O., Avci, O., Kiranyaz, S., Gabbouj, M., Inman, D.J.: Real-time vibration-based structural damage detection using one-dimensional convolutional neural networks. J. Sound Vib. **388**, 154–170 (2017)
11. Zhao, R., Yan, R., Chen, Z., Mao, K., Wang, P., Gao, R.X.: Deep learning and its applications to machine health monitoring. Mech. Syst. Signal Process. **115**, 213–237 (2019)
12. Seventekidis, P., Giagopoulos, D., Arailopoulos, A., Markogiannaki, O.: Structural health monitoring using deep learning with optimal finite element model generated data. Mech Syst Signal Process. **145**, 106972 (2020)
13. Giagopoulos, D., Arailopoulos, A.: Computational framework for model updating of large scale linear and nonlinear finite element models using state of the art evolution strategy. Comput. Struct. **192**, 210–232 (2017)
14. Giagopoulos, D., Arailopoulos, A., Natsiavas, S.: A Model-Based Fatigue Damage Estimation Framework of Large Scale Structural Systems, Structural Health Monitoring (SHM), Article in Press (2019)
15. Wu, J., Luo, Z., Zhang, N., Zhang, Y., Walker, P.D.: Uncertain dynamic analysis for rigid-flexible mechanisms with random geometry and material properties. Mech. Syst. Signal Process. **85**, 487–511 (2017)
16. Daouk, S., Louf, F., Dorival, O., Champaney, L., Audebert, S.: Uncertainties in structural dynamics: overview and comparative analysis of methods. Mech. Ind. **16**(4), 2–3 (2015)
17. Sun, R.-B., Yang, Z.-B., Yang, L.-D., Qiao, B.-J., Chen, X.-F., Gryllias, K.: Planetary gearbox spectral modeling based on the hybrid method of dynamics and LSTM. Mech. Syst. Signal Process. **138**, 106611 (2020)
18. Ma, R., Chen, Y., Cao, Q.: Research on dynamics and fault mechanism of spur gear pair with spalling defect. J. Sound Vib. **331**(9), 2097–2109 (2012)
19. Yang, X., Ding, K., He, G.: Accurate separation of amplitude-modulation and phase-modulation signal and its application to gear fault diagnosis. J. Sound Vib. **452**, 34–50 (2019)
20. Kar, C., Mohanty, A.R.: Vibration and current transient monitoring for gearbox fault detection using multiresolution Fourier transform. J. Sound Vib. **311**(1–2), 109–113 (2008)

Chapter 7
Structural Damage Detection Framework Using Metaheuristic Algorithms and Optimal Finite Element Modeling

Ilias Zacharakis and Dimitrios Giagopoulos

Abstract The vibration-based approach is a subcategory of SHM methods that rely on the fact that a structural damage will affect the dynamic characteristic of a structure. The recent trends show that there is an increasing interest in the use of machine learning (ML) for SHM systems that rely on the experimentally measured data or artificially collected data to properly train the ML model for classification. The proposed method however is taking another approach which is based on the experimental vibration measurements and FE models. However, there is no need to acquire large datasets, and the only requirement is the development of an optimal FE model which has a high correlation with the dynamic response of the physical structure. The optimal FE model of the healthy structure is developed using appropriate FE model updating techniques and experimental vibration measurements, simulating the undamaged condition. A parametric area is inserted into the FE model, changing stiffness and mas to simulate the effect of the physical damage. This area is controlled by the metaheuristic optimization algorithm, which is embedded in the proposed Damage Detection Framework. For effective damage localization, the Transmittance Functions from acceleration measurements are used which have shown to be sensitive to structural damage while requiring output-only information. Finally, with proper selection of the objective function, the error that arises from modeling a physical damage with a linear damaged FE model can be minimized thus creating a more accurate prediction for the damage location. The effectiveness of the proposed SHM method is demonstrated on a CFRP composite beam structure. In order to check the robustness of the proposed method, two small damage scenarios are examined for each validation model and combined with random excitations.

Keywords Damage detection · Vibration · FEM updating · Metaheuristic · Model-based

7.1 Introduction

Reliable and accurate SHM systems are a matter of interest through a wide variety of engineering disciplines such as civil, mechanical, and aerospace. This interest can be attributed to economical, time-saving factors but also the prevention of fatal accidents. The vibration-based methodologies are continuously attracting attention [1, 2].

The presented vibration-based methodology relies on FE model updating techniques and optimization algorithms in order to approximate the location of the damage. Similar approaches have been developed in the past. The majority of the previously developed approaches relied on simplified FE models with most of them using beam elements. Furthermore, a large number of optimization parameters were used, such as the stiffness of each element, to locate the damage in truss-like structures [3–7]. This can lead to changing the global stiffness and/or mass matrix. The proposed method is following a different approach by changing only a submatrix in order to locate the damaged area. This is achieved by inserting a parametric area into the FE model, changing stiffness and mass to simulate the effect of the real damage in physical structure. This area is controlled by a metaheuristic optimization algorithm, in this case the Particle Swarn Optimization algorithm [8]. The presented Damage Detection Framework is using a fixed number of six optimization parameters, while they can be reduced depending on the geometry of the structure. Furthermore, the presented approach is utilizing response-only information from the structure by using the Transmittance Function [9].

I. Zacharakis · D. Giagopoulos (✉)
Department of Mechanical Engineering, University of Western Macedonia, Kozani, Greece
e-mail: izacharakis@uowm.gr; dgiagopoulos@uowm.gr

R. Madarshahian, F. Hemez (eds.), *Data Science in Engineering, Volume 9*, Conference Proceedings of the Society for Experimental Mechanics Series, https://doi.org/10.1007/978-3-031-04122-8_7

For validation purposes, the proposed Damage Detection Framework is tested on an experimental CFRP beam. Two different damage cases are evaluated: the first is an added mass scenario where the second is a local stiffness reduction. In both cases, the damaged area was identified successfully.

7.2 Background

A valid concern in any model-based SHM method is the accuracy of the FE model to simulate the dynamic response of the real structure. Using experimental measurements from the healthy state of the structure and FE model updating procedures, this drawback can be surpassed. In order to acquire an optimal FE model, the covariance matrix adaptation–evolutionary strategy (CMA-ES) optimization algorithm is employed which has been used successfully in the past with linear and nonlinear FE models [10, 11].

The proposed Damage Detection Framework can be applied after the development of the Optimal Finite Element Model. The core of the Framework is an another metaheuristic algorithm, the Particle Swarn Optimization (PSO) [8, 12]. It is a population-based algorithm that belongs to the subarea of Swarm Intelligence in the Computational Intelligence category.

Given a real-world structure ,$\mathcal{S}$, with the corresponding optimal FE model, $\mathcal{M}$, their relationship can be described as :

$$\mathcal{M} = \mathcal{S} + e_1 \tag{7.1}$$

When a damage occurs at the real-world structure, Eq. (7.1) can be transformed as:

$$\mathcal{M}_{\text{dam}} = \mathcal{S}_{\text{dam}} + e_2 \Rightarrow \mathcal{M} + d\mathcal{M} = \mathcal{S} + d\mathcal{S} + e_2 \tag{7.2}$$

where $\mathcal{S}_{\text{dam}}$ and $\mathcal{M}_{\text{dam}}$ represent the damaged real-world structure and the corresponding damaged FE model that approximates the response of the structure as close as possible. The parameters e_1, e_2 represent the error of the model and the physical structure. The proposed Damage Detection Framework is using the optimization algorithm, properly configured in order to find the $d\mathcal{M}$ and thus the $\mathcal{M}_{\text{dam}}$.

The Optimal Finite Element Model ($\mathcal{M}$), of the examined structure, can be described by the equation of motion:

$$\mathbf{M}\,\ddot{x} + \mathbf{C}\,\dot{x} + \mathbf{K}\,x = \mathbf{F} \tag{7.3}$$

where **F** is the external excitation and $\ddot{x}$, $\dot{x}$, x represents the acceleration, velocity, and displacement vectors. The matrices **M**, **C**, **K** are the mass, damping, and stiffness matrix accordingly, and so the model can be fully described by these matrices, $\mathcal{M}$ (M, C, K). Assuming that a damage in a structure will affect the mass and stiffness matrices, the major target is to find the appropriate $d\mathbf{M}$ and $d\mathbf{K}$ that results on the corresponding $d\mathcal{M}$ and so the $\mathcal{M}_{\text{dam}}$ that approximates the damaged structure $\mathcal{S}_{\text{dam}}$.

To achieve this goal, a parametric area is inserted into the FE model. The location and mechanical properties of the area are controlled by the optimization algorithm. Thus, the optimization search domain includes a total of six parameters; two of them represent the percentage change of the Elastic Modulus and Density. The other four parameters represent the location of the inserted damaged area. As such one parameter is assigned for the selection of the affected part, in a multipart structure, and the other three are the coordinates of the damaged area inside the part.

The Transmittance Functions (TF) are used as a metric of comparison between the FE model and the experimental structure. It is a response-only metric computed only between output acceleration signals and is defined from Eq. (7.4).

$$T_{rs}(\omega) = \frac{S_{rs}(\omega)}{S_{rr}(\omega)} = \frac{\ddot{x}_r(\omega)\,\ddot{x}_s^*(\omega)}{\ddot{x}_r(\omega)\,\ddot{x}_r^*(\omega)} \tag{7.4}$$

For instance, $\ddot{x}(\omega)$ is the Fourier transform of the acceleration signal with ω to be the frequency. Furthermore, the $\ddot{x}^*(\omega)$ is the complex conjugate of $\ddot{x}(\omega)$, and subscripts r, s denote the degrees of freedom on the structure.

In order to successfully apply the Framework, the objective function was formed from the Pearson correlation coefficient which takes values from -1 to 1. Furthermore, as the value 1 represent a perfect match between two data sets, the mean value of all the Pearson distances, Eq. (7.5), was set as the objective function for minimization.

$$\text{Pearson Distance} = 1 - \rho(A, B) \tag{7.5}$$

7.3 Analysis

The experimental setup consists of a circular hollow section CFRP composite beam, while two custom-made aluminum connectors were glued at the sides of the beam. One connector was fixed and the other was mounted with an electrodynamic shaker, while two triaxial accelerometers (A1, A2) were placed along the length of the beam. The corresponding FE model was developed which, together with the complete experimental setup, is presented in Fig. 7.1.

Experimental measurements were executed on the healthy state of the structure in order to develop the optimal FE model using the CMA-ES algorithm. Table 7.1 presents the natural frequencies of the physical structure of the nominal FE model and the optimal FE model. Furthermore, Fig. 7.2 shows the comparison of the Transmittance Functions and the FE models.

On the current experimental setup, two damage cases are examined. The first includes an added mass along the beam, while the second is a local stiffness reduction caused by small local cracks on the matrix of the composite. The stiffness was caused by placing the beam at a three-point bending machine using a sharp angle head. In parallel, protective rubber was placed at the other two connection points of the tube with the bending machine to prevent any unwanted damage. In both cases, the location of the damage was the same at 700 mm along the length of the beam. The damaged cases are presented in Fig. 7.3.

After the acquisition of experimental measurements from both damage cases, the Damage Detection Framework was applied. In both cases, it was able to approximate the damaged area correctly. Figures 7.4 and 7.5 present the final damaged FE model created by the Damage Detection Framework. The red highlighted area is indicating the damaged area which was inserted by the Framework that includes the damaged area of the real structure. Furthermore, Fig. 7.6 presents the comparison of Case 2 damage between the Transmittance Functions. It includes the TFs of the healthy and damaged experimental structures but also the TF of the damaged FE model as a result of the Damage Detection Framework.

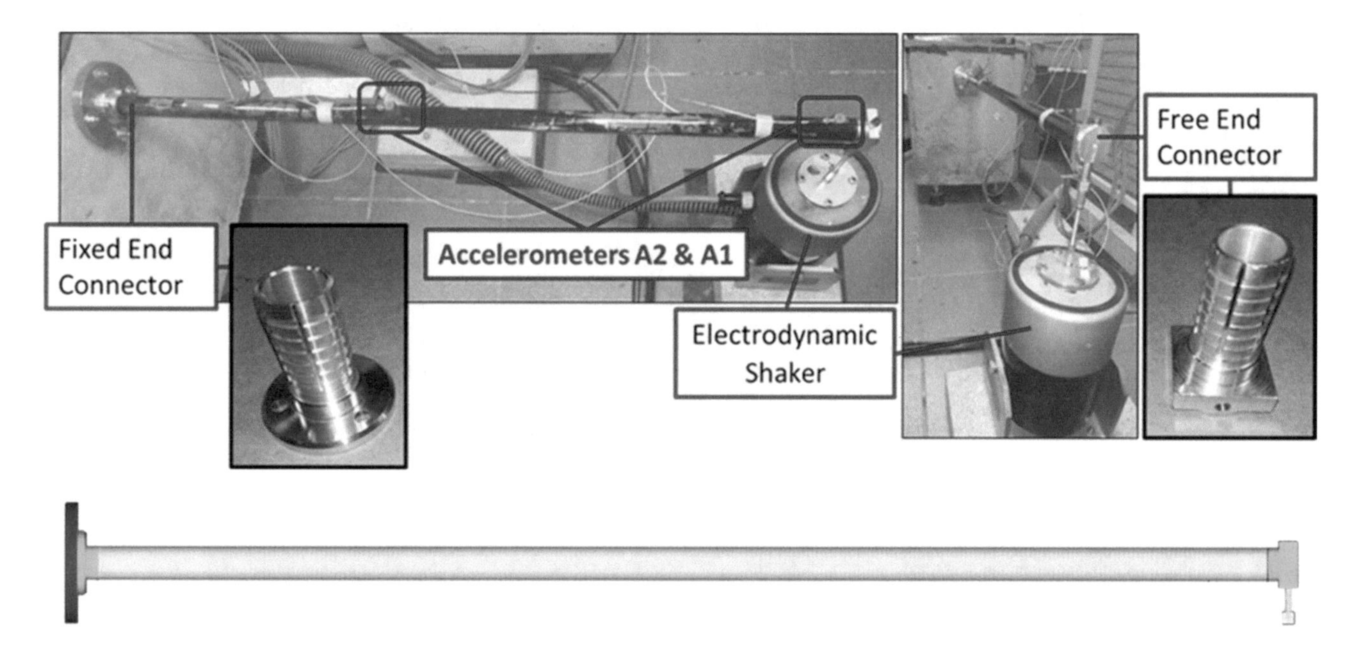

Fig. 7.1 Complete experimental setup of the CFRP composite beam along with the corresponding FE model

Table 7.1 Comparison of the experimental natural frequencies with the Nominal and Optimal Finite Element Model

Mode	Experimental freq. (Hz)	Nominal FE freq. (Hz)	Optimal FE freq. (Hz)
1	136.0	178.6	138.6
2	169.0	210.5	170.4
3	255.0	341.3	260.4
4	476.6	587.2	485.1
5	503.4	600.0	506.5

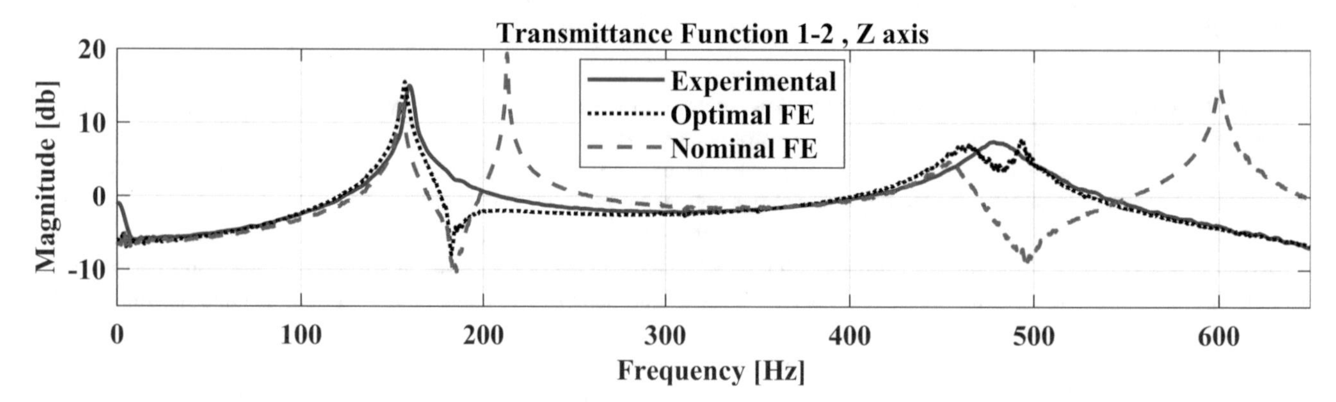

Fig. 7.2 Comparison of the Transmittance Functions 1–2 between the experimental measurements, the nominal FE model and the optimal FE model

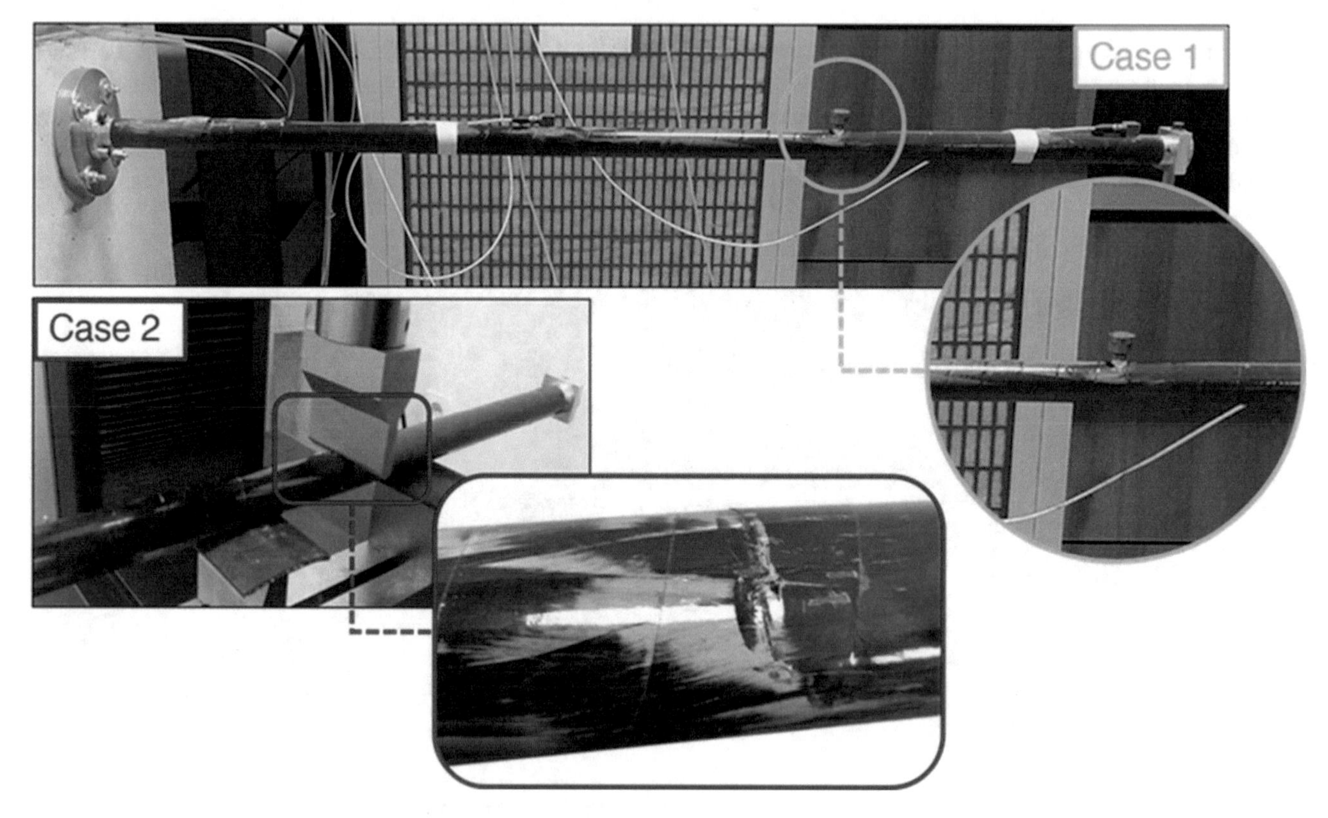

Fig. 7.3 CFRP composite beam damage cases

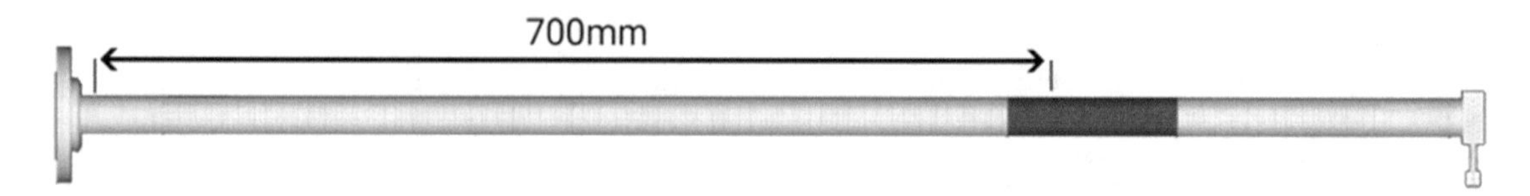

Fig. 7.4 Damaged FE model. Case 1 mass damage

7.4 Conclusion

In this work, a vibration-based Damage Detection Framework was presented using output-only experimental information and embedding FE model updating techniques with metaheuristic optimization algorithms. The presented Framework has the ability to locate the damaged area in a structure. This is achieved by approximating the dynamic behavior of the damaged

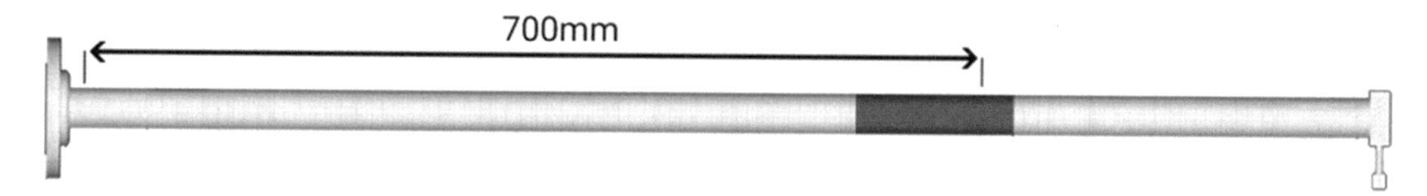

Fig. 7.5 Damaged FE model. Case 2 crack damage

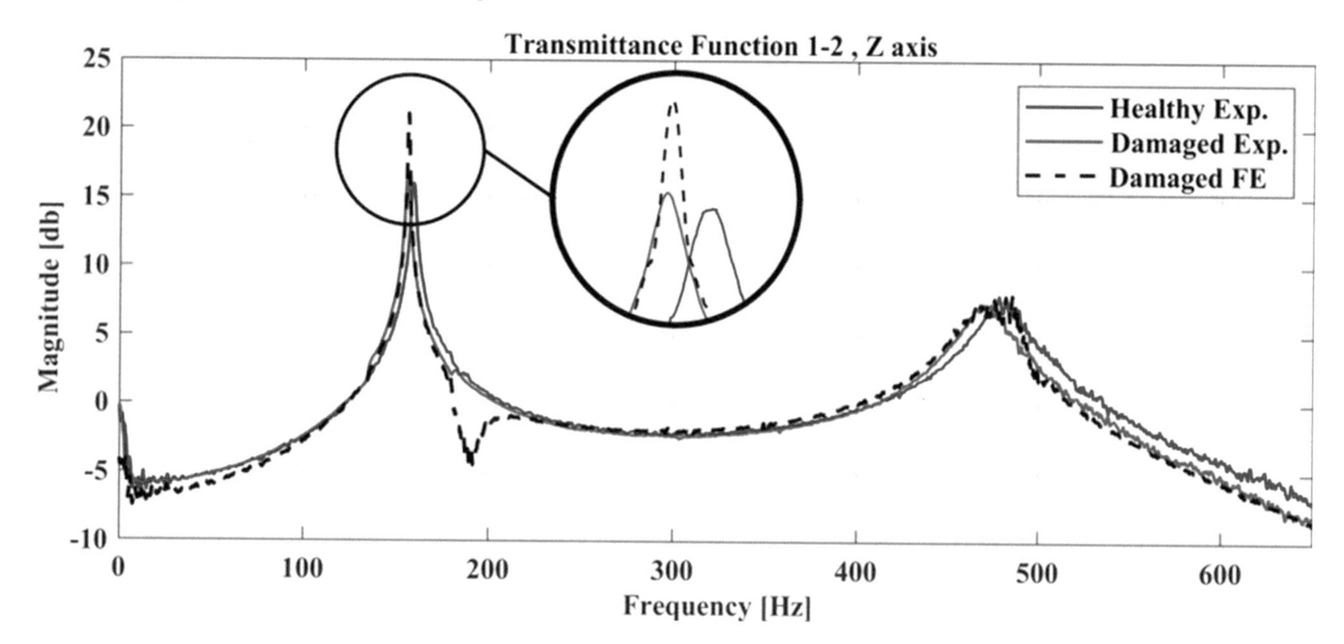

Fig. 7.6 Case 2. Comparison of the Transmittance Functions 1–2 between the damaged and healthy experimental measurements along with the damaged FE model

structure with the use of a linear FE model. A key component of the procedure is the development of an optimal FE model. In order to locate the affected area, a parametric damaged area is inserted at the optimal FE model changing stiffness and mass to approximate the effect of the physical damage. The location of the damaged area on the FE model is controlled by the optimization algorithm. When convergence is achieved, the final location will approximate the location of the physical damage. In order to compare the physical structure and the FE model, the Transmittance Functions were employed that were derived from the acceleration signals and have shown to be prone to material damages. The effectiveness of the proposed Framework is demonstrated by a real experiment on a CFRP composite beam setup using two different damage cases.

Acknowledgments This research has been cofinanced by the European Regional Development Fund of the European Union and Greek national funds through the Operational Program Competitiveness, Entrepreneurship, and Innovation, under the call RESEARCH – CREATE – INNOVATE (project code: T1EDK:05393).

References

1. Doebling, S.W., Farrar, C.R., Prime, M.B.: A summary review of vibration-based damage identification methods. Shock Vib. Dig. **30**, 91–105 (1998). https://doi.org/10.1177/058310249803000201
2. Gomes, G.F., Mendez, Y.A.D., Alexandrino, P.S.L., da Cunha, S.S., Ancelotti, A.C.: A review of vibration based inverse methods for damage detection and identification in mechanical structures using optimization algorithms and ANN. Arch. Comput. Methods Eng. **26**, 883–897 (2019). https://doi.org/10.1007/s11831-018-9273-4
3. Jafarkhani, R., Masri, S.F.: Finite element model updating using evolutionary strategy for damage detection. Comput. Civ. Infrastruct. Eng. **26**, 207–224 (2011). https://doi.org/10.1111/j.1467-8667.2010.00687.x
4. Friswell, M.I.: Damage identification using inverse methods. Philos. Trans. R. Soc. A Math. Phys. Eng. Sci. **365**, 393–410 (2007). https://doi.org/10.1098/rsta.2006.1930
5. Friswel, M., Penny, J.E., Wilson, D.A.L.: Using vibration data and statistical measures to locate damage in structures. Int. J. Anal. Exp. Modal Anal. **9**, 239–254 (1994)

6. Vo-Duy, T., Ho-Huu, V., Dang-Trung, H., Nguyen-Thoi, T.: A two-step approach for damage detection in laminated composite structures using modal strain energy method and an improved differential evolution algorithm. Compos. Struct. **147**, 42–53 (2016). https://doi.org/10.1016/j.compstruct.2016.03.027
7. Ding, Z.H., Huang, M., Lu, Z.R.: Structural damage detection using artificial bee colony algorithm with hybrid search strategy, Swarm. Evol. Comput. **28**, 1–13 (2016). https://doi.org/10.1016/j.swevo.2015.10.010
8. Kennedy, J., Eberhart, R.: Particle swarm optimization. In: Proceeding of the ICNN'95 – International conference on Neural Networks, IEEE, 1995, pp. 1942–1948. https://doi.org/10.1109/ICNN.1995.488968.
9. Zhang, H., Schulz, M.J., Naser, A., Ferguson, F., Pai, P.F.: Structural health monitoring using transmittance functions. Mech. Syst. Signal Process. **13**, 765–787 (1999). https://doi.org/10.1006/mssp.1999.1228
10. Giagopoulos, D., Arailopoulos, A.: Computational framework for model updating of large scale linear and nonlinear finite element models using state of the art evolution strategy. Comput. Struct. **192**, 210–232 (2017). https://doi.org/10.1016/j.compstruc.2017.07.004
11. Zacharakis, I., Giagopoulos, D., Zyganitidis, I., Arailopoulos, A., Markogiannaki, O.: Modeling of cfrp structures using model updating techniques and experimental measurements. In: Proceedings of the international conference on Structural Dynamics, EURODYN, 2020: pp 536–550. https://doi.org/10.47964/1120.9042.19310
12. Eberhart, R., Kennedy, J., New optimizer using particle swarm theory. In: Proceedings of the international symposium on micro machine and human science (1995) pp 39–43. https://doi.org/10.1109/mhs.1995.494215.

Chapter 8
On Aspects of Geometry in SHM and Population-Based SHM

Chandula T. Wickramarachchi, Jack Poole, Elizabeth J. Cross, and Keith Worden

Abstract Population-based structural health monitoring (PBSHM) allows inferences to be made between and within groups of structures by transferring knowledge across them. In order to successfully apply PBSHM, similarity within the populations needs to be assessed. In the PBSHM framework, structural similarity can be evaluated using metrics on irreducible element models and attributed graphs, while similarity in data can be determined using a number of information/distance metrics. In this chapter, the need for metrics to measure similarity in structures and data within PBSHM is discussed using ideas of geometry, from simple rigid transformations to fibre bundles. The main aim of the chapter is to consider similarity in data using distance metrics with a special focus on transfer learning and data standardisation/normalisation.

Keywords PBSHM · Similarity metrics · Transfer learning · Standardisation/normalisation

8.1 Introduction

In the recent past, certain serious problems in the implementation of Structural Health Monitoring (SHM), have motivated the development of a new variant—*population-based SHM* (PBSHM) [1–3]. The main aim of the theory is to allow the transfer of health data, information and inferences, between different structures in a specified population. In some cases, this is critical if one wishes to move to higher-level diagnostics than simple novelty detection. Such high-level classifiers require data from all relevant damage states of the structure of interest, and these data may not be available; it is often the case that only normal-condition data are available. PBSHM attempts to solve the problem by exploiting damage-state data from *similar* structures in the population. If such data can be found, one can use *transfer learning* to improve or extend SHM capability on the original structure [3]. The crucial word here is *similar*; transfer learning will only work between similar learning problems, and in the SHM context this translates to similar *structures*. If transfer is attempted between dissimilar problems, there is the serious risk of *negative transfer*, a situation in which transfer makes the original capability *worse*.

A core problem in PBSHM is then how to measure the similarity of structures. The general theory adopts quite an abstract approach to this problem [2]. The idea is to create an abstract representation theory for structures, in which structures are represented as points in a *metric space*. The defining property of a metric space is that it is equipped with a measure of distance—the metric [4]. Suppose that points in the space of structures S are denoted by a vector $\underline{x}$, a *metric* $d(\underline{x}, \underline{y})$ is a function of two points in S, with certain properties: (i) $d(\underline{x}, \underline{y}) \geq 0$, with $d(\underline{x}, \underline{y}) = 0$ iff (if and only if) $\underline{x} = \underline{y}$; (ii) $d(\underline{x}, \underline{y}) = d(\underline{y}, \underline{x})$ (symmetry); $d(\underline{x}, \underline{y}) \geq d(\underline{x}, \underline{z}) + d(\underline{z}, \underline{y})$ (triangle inequality). If S were an ordinary real vector space, the Euclidean distance $d(\underline{x}, \underline{y}) = \sqrt{(\underline{x} - \underline{y})^T(\underline{x} - \underline{y})}$ would be the standard metric. If the space of structures S in PBSHM can be created/defined as a metric space, then the metric will serve as the required measure of similarity of structures.

Metrics have been essential in data-based SHM since the inception of the subject [5]; they are the basis of all novelty detection. The Mahalanobis squared-distance, used to detect outliers [6], is a clear example. In this case, the metric operates on the *feature space* associated with health-state data. Such metrics will still be important in PBSHM as diagnostic tools; so two types of space with their corresponding metrics are already singled out. Metrics are an important ingredient in the mathematical subject of *geometry* [7], so it seems that—in abstract terms—PBSHM appears to be a very geometrical theory. This observation is discussed in some detail in [8], where many aspects of PBSHM are considered in terms of *fibre bundles*.

C. T. Wickramarachchi (✉) · J. Poole · E. J. Cross · K. Worden
Dynamics Research Group, Department of Mechanical Engineering, University of Sheffield, Sheffield, UK
e-mail: c.t.wickramarachchi@sheffield.ac.uk; jpoole4@sheffield.ac.uk; e.j.cross@sheffield.ac.uk; k.worden@sheffield.ac.uk

R. Madarshahian, F. Hemez (eds.), *Data Science in Engineering, Volume 9*, Conference Proceedings of the Society for Experimental Mechanics Series, https://doi.org/10.1007/978-3-031-04122-8_8

While geometry is fundamentally concerned with the properties of spaces, *transformations* on those spaces are also critical ingredients [9]. Of course, transformations of feature spaces are a fundamental tool in SHM; for example, principal component analysis (PCA) is nothing more than a rigid rotation followed by an optional (but usual) projection. It is therefore natural to think about how transformations might be a fundamental ingredient of PBSHM; it will be shown later that this viewpoint leads to useful ideas about 'normalisation' of data.

This chapter will be composed of a number of observations about how geometry appears to be the fundamental guiding principle for PBSHM. The layout of the chapter is as follows. The next section will discuss the fibre-bundle viewpoint on PBSHM; this is followed by sections discussing the use of metrics in SHM and PSHM. Section 8.5 then presents how groups of transformations can help prepare feature spaces for more effective transfer learning. The final section presents some general conclusions.

8.2 Fibre Bundles for PBSHM

In the chapter [8], the fundamental objects of PBSHM are unified in a single geometrical structure—a *fibre bundle*. The bundle allows a pictorial representation of all the main concepts in PBSHM, the relevant spaces and their metrics, as in Fig. 8.1.

The fibre bundle is composed of two spaces: the space of structures S (the base space) and the total space T. The other crucial ingredient in the bundle is a projection map $\pi : T \longrightarrow S$. The inverse of π will usually be multivalued, and the sets $\pi^{-1}(s)$ in T are referred to as the *fibres* above the points $s \in S$. For a bundle, it is required that all the fibres are *homeomorphic* to a single space F (essentially 'all the same' space). In PBSHM, the fibre above a structure s will be identified with the *feature space* of that structure. A map $f : S \longrightarrow T$, which chooses a single point in the fibre for each base point, is called a *section*. Clearly, sections have the property that the composition $\pi \circ f$ is the identity. Sections are essentially *fields* over the base space. In PBSHM, the base space S is the space of *attributed graphs* that represent structures [2]; this is quite a complicated space; in particular, it is not a differentiable manifold and does not support 'traditional' calculus. The fibres F,

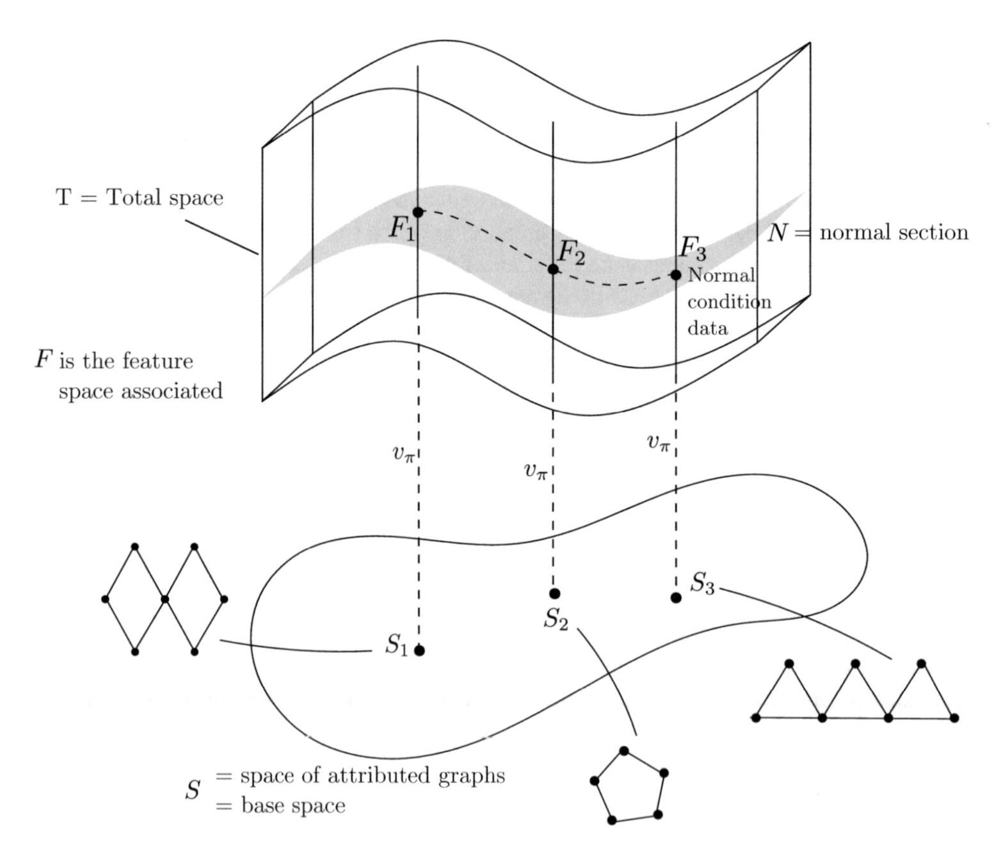

Fig. 8.1 Schematic showing the fibre bundle structure of PBSHM

on the other hand, are usually quite simple spaces; for example, if the features for each structure are the first four natural frequencies of the structure, F is simply $\mathbb{R}^4$—a four-dimensional real vector space.

A fundamental object in PBSHM is the normal section n, which is the map that associates with each structure its normal-condition feature data (associated with some standard conditions on any environmental or operation variables; e.g. standard temperature and pressure). Any section represents an embedding of the base space S in the total space T. An interesting consequence of this is that metrics on T induce metrics on S; that is given a metric d_T on T, and a section f, a metric on S can be defined by $d_S(S_1, S_2) = d_T(f(S_1), f(S_2))$; this is a type of *pull-back* [9]. The implication for PBSHM is that 'similarity in data implies similarity in structure'; this is the converse of the hypothesis for positive transfer, which is 'similarity in structure implies similarity in data'.

A much more detailed discussion of the bundle structure of PBSHM can be found in [8]; for now, attention will be concentrated on the metrics that are fundamental to the theory.

8.3 Metrics in SHM and PBSHM

8.3.1 Metrics on the Space of Structures

As discussed in [2], the space of structures S in PBSHM is a space of *attributed graphs* (AG). These graphs are constructed by first identifying a reduced physical model of the structure of interest called an *irreducible element* (IE) model. Mathematically, a graph is simply a collection of *vertices* (or nodes) V, which are connected by some set of edges E; a given graph is thus denoted as $G(V, E)$. Physically, the vertices in a structural graph represent the major components or substructures and the edges represent the joints between these objects; this means that the graph captures the overall topology of the structure. In an attributed graph, vectors of parameters can be attached to nodes or edges; for example, one could attach the physical dimensions of a structural component to its node in the graph.

As discussed in the introduction, a basic problem in PBSHM is to assess the similarity of two structures to decide if transfer learning will be possible between them. Suppose one has a problem with a *target* structure S_t, the idea will be to find the closest *source* structure S_s. If one is equipped with a metric d on the space of structures S, populated by individuals $\{S_1, \ldots, S_P\}$, the problem becomes one of optimisation:

$$S_s = \underset{i}{\operatorname{argmin}} \; d(S_i, S_t) \tag{8.1}$$

At a more sophisticated level, one might use the k closest structures to S_t and then use *multi-task learning* rather than transfer learning [10].

The important question here relates to the choice of the metric d on the space of graphs. In the literature, such metrics are often discussed in terms of *graph matching*.

Graph Matching

The graph-matching techniques used so far in PBSHM use the idea of the *maximum common subgraph* (MCS). The MCS is the largest common subgraph between two graphs. By assessing the size of the MCS of two graphs, it is possible to infer the similarity of the graphs in question; furthermore, the MCS will correspond physically to the largest common substructure between the corresponding structures [2]. To find the MCS, the modified Bron–Kerbosch algorithm can be used [11]. Once the MCS is found, similarity metrics can be used on the subgraphs, which can also take account of any attributes.

If the two graphs to be compared are of the same size (same number of nodes), the metrics that can be used fall into two categories: the *spectral* and *matrix* distances [12], although the authors of that reference suggest that some metrics do not fit neatly into either. To find the spectral distances, a combination of the graph adjacency matrix

$$[A]_{ij} = \begin{cases} 1 & \text{if } (v_i, v_j) \in E \\ 0 & \text{otherwise} \end{cases} \tag{8.2}$$

and the degree matrix (the diagonal matrix of degrees, d_i, counting the number of nodes connected to node i)

$$[D]_{ij} = \begin{cases} d_i & \text{if } i = j \\ 0 & \text{otherwise} \end{cases} \tag{8.3}$$

are used to find the Laplacian matrix $L = D - A$; the *spectrum* of a graph is then defined as the sorted sequence of eigenvalues of L. If the adjacency spectra of two graphs G and G' are λ_i^G and $\lambda_i^{G'}$, respectively, then the *adjacency spectral distance* in the ℓ_2-norm would be

$$d_A(G, G') = \sqrt{\sum_{i=1}^{n} (\lambda_i^G - \lambda_i^{G'})^2} \tag{8.4}$$

The *matrix distance*, on the other hand, first measures the distances within a graph $\delta : V \times V \to \mathbb{R}$ followed by a matrix of pairwise distances $M_{i,j} = \delta(i, j)$. Then the distance induced by δ on G and G' is

$$d(G, G') = \| M - M' \| \tag{8.5}$$

using an appropriate matrix norm.

It will be the usual case in PBSHM that the graphs are not of the same size. By considering the isometry classes of metric spaces, distance metrics can be found between two graphs of different sizes [9]. Gromov–Hausdorff distance, the Kantorovich–Rubinstein distance and the Wasserstein distance are some of the example metrics that can be used in this case.

A much simpler approach to this problem is by using the *Jaccard similarity score* J [2]. By counting the number of vertices in the graphs G and G', and in the MCS H of the two graphs, the Jaccard score given by:

$$J(G, G') = \frac{|V(H)|}{|V(G)| + |V(G')| - |V(H)|} \tag{8.6}$$

can be calculated, where $|.|$ denotes cardinality. The distance between the graphs can then be calculated by $d = 1 - J$. The Jaccard index gives an indication of the topological similarity of two graphs. Although the Jaccard score is rather crude compared to some of the other measures cited here, it has proved remarkably useful for PBSHM already; in [13], it is able to successfully pair bridges of similar type in a population of eight bridges.

The metrics discussed above are completely graph theoretical and only consider topology in the match. It may, however, prove optimal to include more physics into the similarity formulation. Suppose, for example, that the IE/AG model contains enough attributes (density and volume that describe the mass, and dimensions and Young's modulus that indicate the stiffness), to construct a canonical modal model via the lumped masses $[m_{ij}]$ and stiffnesses $[k_{ij}]$ for structures S_i. In this case, one can construct vectors of natural frequencies ω_i and modeshape matrices $[\mathcal{X}_i]$ from s_i. If one then truncates one model (if necessary), so that both have the same number of modes, one could stack ω_i and $[\mathcal{X}_i]$ into a vector $\boldsymbol{\Psi}_i$ for each structures. Finally, a standard Euclidean norm $\| \cdot \|$ can give a measure of similarity between structures S_i and S_j

$$d(S_i, S_j) = \| \Psi_i - \Psi_j \| \tag{8.7}$$

Another principled method for including graph attributes into the similarity measurements is the formulation of graph kernels, which will be discussed next.

Graph Kernels

Unsurprisingly, graph kernels use kernel-based methods to measure similarity within graphs. By using a combination of different positive-definite kernels, graph kernels have the ability to not only compare the topology of the graphs but also include their attributes, encoding the physics and, therefore, the dynamical behaviour of the corresponding structures.

There are many graph kernels available for use, such as *random-walk kernels* [14, 15], *shortest-path kernels* [16], the *GraphHopper kernel* [17] and *Graph-invariant kernels* [18], to name a few. Each of these kernels combinatorially compares the nodes and edges to measure the similarity of the graphs. When comparing the nodes and edges, these methods use a combination of Dirac kernels to assess their types, Gaussian kernels to measure their vectorised attributes and Brownian-bridge kernels to compare their sizes.

The graph kernels themselves are therefore metrics that are capable of assessing the similarity in a population. However, it is also possible to use a distance metric such as the maximum mean discrepancy (MMD) on the resultant kernels [19] for similarity measurement. The MMD is discussed in more detail later on in this chapter.

As discussed previously, the hypothesis from the fibre bundles is that positive transfer is likely if there is similarity between the structures; this then suggests that there is similarity in the collected features. In the next section, the metrics that can be used to measure similarity in the space of features are discussed.

8.3.2 *Metrics on the Space of Features*

'Metrics' have been used in feature spaces throughout the short history of data-based SHM; in fact, most novelty detection methods share some of the properties of metrics, in measuring the 'distance' in some sense from some set representing normal condition. In some ways, the failure to be a true metric does not matter; for example, the Kullback-Liebler divergence is an effective novelty detector but fails to be a true metric because it is not symmetric in its arguments. Where a true metric does exist on the feature spaces, it can be extended to a metric on the total space of the bundle T and can then be 'pulled-back' along a section, to give a metric on the space of structures S. As discussed in Sect. 8.2, the implication is 'similarity in data implies similarity in structures'.

Different metrics on F (and thus T) will induce different metrics on S. One may well want to use different metrics on F. In general, the feature data might actually live on a low-dimensional curved manifold embedded in some flat fibre $F = \mathbb{R}^n$. If that manifold is 'folded' in the embedding space, the Euclidean metric on the space will be deceptive. In that case, one has two options: (a) make the actual fibre F into the real data manifold and use the Riemannian metric on manifold;[1] (b) determine the Riemannian metric on the manifold induced by the embedding (the second fundamental form in geometric terms [9]). In both cases, the more 'true' metric will still induce a metric on the base space of graphs/structures, but it will be different from that induced by the Euclidean metric on the flat fibre.

The use of distance metrics, and the effect of transformation on the space of features will be discussed further here, first with a focus on a single structure, which will then be expanded to a population of structures. Later, metrics used for transfer learning will be explored, while paying attention to data transformations that may help transfer.

Distance Metrics in SHM

First, consider the need for metrics in an SHM problem where data from a single structure is the concern. Assume that data from the undamaged state of a system are in the form of a feature vector $\underline{x} \in \mathbb{R}^n$, with a probability density function $p(\underline{x})$. By convention here, the cumulative density function will be denoted as upper case, so that $P(\underline{x}) = \int p(\underline{x})d\underline{x}$. For illustrative purposes, it will be assumed that p is normally distributed with mean $\underline{\mu}_1$ and covariance matrix Σ_1, i.e. $p = \mathcal{N}(\underline{\mu}_1, \Sigma_1)$. The damage-state data are assumed to be distributed as $q \sim \mathcal{N}(\underline{\mu}_2, \Sigma_2)$. Furthermore, it will be assumed that the data are separable; in practical terms, this means that a measurement from the undamaged state of the structure of interest will be distinguishable from a measurement from the damage state.

The *support* S_p of a given Gaussian distribution p will be defined here as the set with centroid $\underline{\mu}$ such that $\int_{S_p} p(\underline{x})d\underline{x} = 0.999$. S_p and S_q as defined above are ellipses, and if a point is sampled from $p(\underline{x})$, only 1 in 1000 on average, will fall outside S_p. It is thus reasonable to say that the SHM data are separable if $S_p \cap S_q = \phi$ the empty set. Figure 8.2 pictures this situation in $\mathbb{R}^2$. The 'lines' a_i and b_i are the major and minor axes of the support ellipses, which are the principal axes for the densities and can be determined by principal component analysis.

Now, a *classifier* for this problem is a function $f : \mathbb{R}^n - L$, where L is a discrete label space $\{U, D\}$. The function f is learnt from training data $D_u = \{\underline{x}_i^u; i = 1, \dots, N_u\}$, and $D_d = \{\underline{x}_i^d; i = 1, \dots, N_d\}$.

For the first SHM problem, suppose that $D_d = \phi$, i.e. there are no damage-state data. It is possible to frame this as a novelty detection problem. The simplest strategy is to say point $\underline{x}^*$ is novel if $\underline{x}^* \notin S_p$ (obviously there is a confidence level implicit in this; here, it is the 99.9% level). If S_p is a sphere, the test for $\underline{x}^* \in S_p$ is trivial, i.e. $\| \underline{x}^* - \underline{\mu}_1 \| \leq r$, where r is the radius of S_p. Therefore a suitable metric here serves as the test for novelty.

[1] This might not work globally, i.e. one might get different fibres over different points in the base space S, and this will not give a true fibre bundle. However, there are generalisations of manifolds that might prove interesting, like varieties and schemes [20].

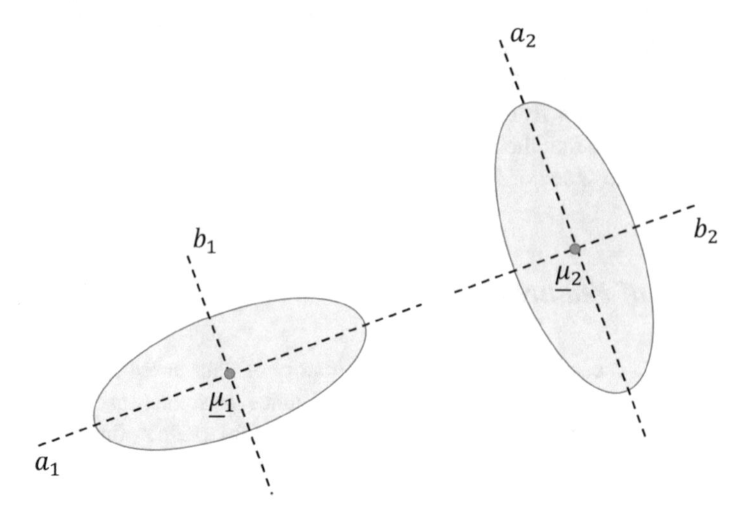

Fig. 8.2 The probability densities of 'undamaged' and 'damaged' data in $\mathbb{R}^2$

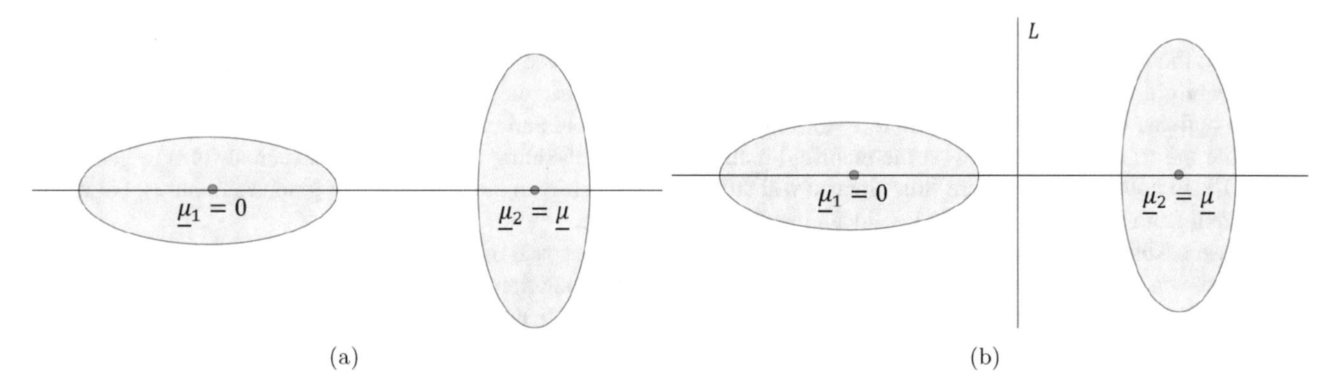

Fig. 8.3 (**a**) Effect of rigid affine transformation on Fig. 8.2. (**b**) The decision boundary of the transformed data

Clearly if S_p is general—i.e. elliptical—this approach does not work. However, using the Mahalanobis–squared distance, $d_M^2 = \| \underline{x}^* - \underline{\mu}_1 \|_\Sigma = (\underline{x}^* - \underline{\mu}_1)^T \Sigma^{-1} (\underline{x}^* - \underline{\mu}_1)$, a simpler test for novelty, $\| \underline{x}^* - \underline{\mu}_1 \| \leq r_M^\alpha$ can be recovered. Here r_M^α is a scalar 'radius'. There is an α-level confidence that $\underline{x}^*$ was drawn from $p(\underline{x})$—and is therefore from the undamaged state—if $\| \underline{x}^* - \underline{\mu}_1 \| \leq r_M^\alpha$.

The novelty detection problem has a naturally occurring error because 0.1% of the time, a normal feature will be assigned to the damage class. In general this error can be reduced by moving the boundary of S_p outwards, but as the location of $\underline{\mu}_2$ is unknown, false-negative errors may occur.

Another 'metric' approach to novelty detection that does not assume a Gaussian distribution is to train an auto-associative neural network A on the D_u such that $\hat{\underline{x}}_i^u = A(\underline{x}_i^u) \approx \underline{x}_i^u \forall \underline{x}_i \in D_u$. Then a test for normal condition is $\| \hat{\underline{x}}^* - \underline{x}^* \| \leq \varepsilon$ for some threshold ε. Usually cross-validation is needed here.

Now suppose there are some damage-state data, i.e. $D_d \neq \phi$; this implies that $\underline{\mu}_2$ and Σ_2 can be estimated. It is now possible to regard this scenario as a true classification problem; treating this as a novelty detection problem will waste information. To aid classification, data can be aligned and transformed in order to rearrange them into a more convenient form (as long as the problem is not affected). In this case, any *rigid affine transformation* $\underline{x} \to A\underline{x} + \underline{b}$ (where A is an invertible matrix and $\underline{b}$ is a vector to represent translation) will be allowed; if A is orthonormal, i.e. $A^T A = I_n$, it will represent a rotation. Here I_n is the $n \times n$ identity matrix. Without loss of generality, a rigid movement on the data can be made so that $\underline{\mu}_1$ is at the origin; that is $\underline{\mu}_1 = 0$ (translation), and $\underline{\mu}_2$ is on the x_1 axis (rotation), leading to $\| \underline{\mu}_2 - \underline{\mu}_1 \| = \underline{\mu}$ (Fig. 8.3a). Note that the covariance matrices will have changed (the principal angles have rotated) but if A is known, the Σs transform like a similarity transformation, $\Sigma \to A^T \Sigma A$.

If novelty detection is being conducted here, it is now possible to reduce the false negatives by moving the threshold away from $\underline{\mu}_1$ as long it is not moved too close to $\underline{\mu}_2$. Ignoring the covariance information, the obvious step is to move the threshold to the midpoint of the line joining $\underline{\mu}_1$ and $\underline{\mu}_2$ (or $\frac{1}{2}\underline{\mu}$) in the aligned problem. Then, $\underline{x}^*$ can be classified as undamaged if $\| \underline{x}^* - \underline{0} \| < \| \underline{x}^* - \underline{\mu} \|$ and vice versa. A decision boundary L can then be defined, which partitions the $\mathbb{R}^n$ into two half-spaces U and D (Fig. 8.3b). If $\underline{x}^* \in D$, then damage can be inferred, etc. When the covariances are accounted for, l becomes a quadratic curve, yet the principle remains the same. This type of reasoning is the basis for the *Fisher discriminant*.

All of the above is based on metrics on the feature space; the same ideas can also be applied in the space of density functions. Suppose the training data are used to make density estimates $\hat{p}(\underline{x})$ and $\hat{q}(\underline{x})$; it is possible to test if a new point $\underline{x}^*$ is more consistent with $\hat{p}$ or $\hat{q}$. Alternatively, a number of $\underline{x}^*$ can be sampled from an unknown condition and a density $\hat{c}(\underline{x}^*)$ could be estimated, where $\hat{c}(\underline{x}^*) \simeq \hat{p}(\underline{x}^*)$ would suggest that the system is in the normal condition.

Metrics can also be useful here. The $\hat{c}$ and $\hat{p}$ live in a function space $\mathcal{F}$. If that space has a metric—say induced by a norm $\| \cdot \|_{\mathcal{F}}$—then the hypothesis test $\hat{c}(\underline{x}) \simeq \hat{p}(\underline{x})$ can be converted to $\| \hat{c} - \hat{p} \|_{\mathcal{F}} \simeq 0$. In fact, a full metric on $\mathcal{F}$ is not needed here, only a monotonically increasing functional, which is a minimum when $\hat{c} = \hat{p}$, is needed. One candidate is the Kullback-Lieber (KL) divergence

$$D_{KL}(p||q) = \int p(\underline{x}) \log \frac{p(\underline{x})}{q(\underline{x})} d\underline{x} \tag{8.8}$$

which is 0 when $p = q$. D_{KL} is not a metric because it is not symmetric, i.e. $D_{KL}(p||q) \neq D_{KL}(q||p)$; however, it can be trivially symmetrised by taking $D_{KL}(p||q) + D_{KL}(q||p)$.

The problem with metrics like the KL-divergence is that they operate on probability density functions. If a simple parametric form is appropriate, like a Gaussian distribution with known mean and covariance, the KL-divergence can be estimated by numerical integration. However, if only samples of data are known, it is necessary to compute density estimates, e.g. via *kernel density estimation* [21]. Unfortunately, density estimation is one of the most difficult learning problems and suffers so greatly from the *curse of dimensionality* [22], which estimates in more than four or five dimensions will not usually be feasible on engineering data sets, without some form of dimension reduction. Furthermore, as there is no universally accepted density estimation method, the choice can cause variability in results. Fortunately, there exist metrics that do not require an explicit density function; such a metric is the *maximum mean discrepancy* (MMD). The MMD works by embedding the sample data in a *reproducing-kernel Hilbert space* $\mathcal{H}$, via a nonlinear map ϕ; the MMD is then computed as the distance between the *means of the embeddings* in the $\mathcal{H}$ metric. If the data from the two distributions are $D_1 = \{\underline{x}_1, \dots \underline{x}_N\}$ and $D_2 = \{\underline{y}_1, \dots \underline{y}_M\}$, then the MMD is defined as [23]

$$\mathrm{MMD}(D_1, D_2) = \left\| \frac{1}{N} \sum_{i=1}^{N} \phi(\underline{x}_i) - \frac{1}{M} \sum_{i=1}^{M} \phi(\underline{y}_i) \right\|_{\mathcal{H}} \tag{8.9}$$

The MMD has the property that $\mathrm{MMD}(p, q) = 0$ if, and only if, $p = q$.

8.4 Metrics and Sections in PBSHM

It is now (hopefully) clear that metrics of various sorts are both explicit and implicit in data-based SHM problems. The discussion will now return to the case of PBSHM and will reference the fibre-bundle framework discussed earlier and depicted in Fig. 8.1. In general, the normal condition data for a structure s_i will be specified by a training set $D_i = \{\underline{x}_j^{(i)} : j = 1, \dots N_i\}$. In this case, the simplest situation would be when the fibre F is a real vector space of some dimension d: $F = \mathbb{R}^d$. The simplest metric on F might then be $d(\underline{x}_j^{(i)}, \underline{x}_k^{(i)}) = \| \underline{x}_j^{(i)} - \underline{x}_k^{(i)} \|$. This is just as it is in an individual SHM problem; a simple extension to PBSHM can be made as follows. If it is assumed that the features are the same for each structure in the population, then the fibres (feature spaces) will all be isomorphic: e.g. if the features are the first four natural frequencies, the fibres would all be $\mathbb{R}^4$. In this case, one can assume that they are the *same* $\mathbb{R}^4$. This is of course, what one does when one plots data from different structures on the same axes. Suppose that the mean features for two structures S_1 and S_2 are $\underline{\mu}_1$ and $\underline{\mu}_2$; then the distance between the structures can be defined as $||\underline{\mu}_1 - \underline{\mu}_2||$. By this identification of the fibres, one can also use metrics on probability distributions to induce metrics on S. There is clearly a subtlety here; if the data from S_1 represent the healthy state of that structure and the data from S_2 represent a damage state, it is clearly senseless to use $||\underline{\mu}_1 - \underline{\mu}_2||$ as a distance between structures. The solution is to ensure that the distance in the feature space is only computed between corresponding classes. The obvious choice of a *corresponding class* is the healthy state for each structure. In this case, $\underline{\mu}_1$ and $\underline{\mu}_2$ are on the normal section of the bundle, and one recovers the distance $d(S_1, S_2) = ||\underline{\mu}_1 - \underline{\mu}_2|| = ||n(S_1) - n(S_2)||$. Thus, one has recovered the *pullback* metric on S from earlier, via a 'common-sense' argument.

The strength of PBSHM, however, is in a multi-class scenario, where some data for a structure of interest are unavailable. In this case, domain adaptation in the form of transfer learning [3, 24, 25] can prove very powerful. In the next section, useful metrics and helpful data transformations for transfer learning are discussed.

8.5 Transformations in Transfer Learning

Transfer learning (TL), in the form of domain adaptation, is a technique whereby labelled and unlabelled data from separate structures are mapped onto the same space. Unlike the mappings just discussed where the feature spaces were identified in a single universal fibre, TL maps *from* this space in such a way that pairs of clusters (corresponding to the same labels—in the case of PBSHM, health-state labels) are harmonised so that a classifier trained on the data from one structure will generalise to another. TL is particularly useful in helping classifiers trained on some *source* domain using labelled data, to generalise well in a *target* domain containing *unlabelled* data [25]. In PBSHM, the target problem will be associated with some structure S_t for which the 'right' data are unavailable, and transfer is conducted from a structure S_s for which suitable data *are* available.

In the PBSHM bundle, there will be a number of candidate structures S_s^i, with source data D_i, that could be used for transfer; however, each structure will carry some threat of *negative transfer*. Negative transfer occurs when the learner performance is *worse* if data or information from any domain other than the target is used [24]. This occurs when the target problem (SHM problem) on S_t is significantly different from the source problem (SHM problem) on S_s; it is expected that this will happen if S_t is a significantly different structure to S_s. The question then is: for which structures S_s^i is transfer advisable [26]? A principled means of finding the structure S_s^{i*} that minimises this threat is needed. A possible solution is already formulated in Eq. (8.1) and will not be discussed further. Instead, it will be interesting to consider how groups of *transformations* can be used to simplify and otherwise aid transfer. Such transformations will be referred to here as *Pre-Transfer-Alignment* (PTA), although it will become clear that PTA helps to harmonise feature spaces itself and could thus be considered a form of transfer.

Consider the situation shown in Fig. 8.4, which represents a schematic of a two-class TL problem in two-dimensional space. For simplicity, all classes will be assumed to be distributed as spherical Gaussians. One can regard Class 1 (the lighter shade in each case), as the normal-condition data for an SHM problem, and Class 2 (darker), as representative of a damage state.

Firstly, without any the loss of generality, the source problem can be aligned (Fig. 8.5); this uses the rotation degree-of-freedom of an affine transformation on the source.

Suppose for now that there are some normal-condition data in the target domain, but no damage-state data. Clearly in PTA, the mean can be removed in the target domain, giving Fig. 8.6a; this uses the translation degree-of-freedom for an affine transformation on the target. There is no basis for making a rotation of the target, because the position of the cluster is

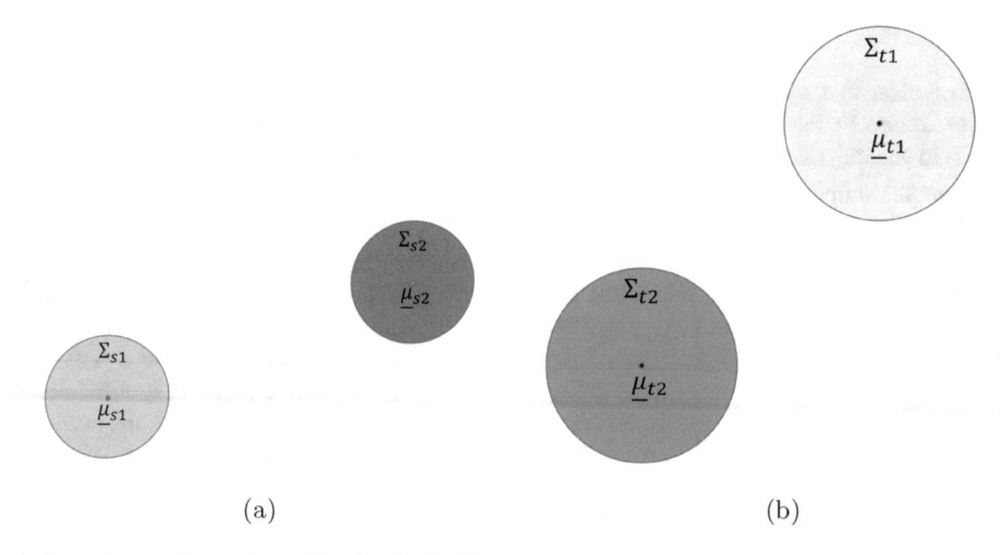

Fig. 8.4 A Source (**a**), and target (**b**), problem of TL visualised in 2D

Fig. 8.5 The source problem aligned according to the mean

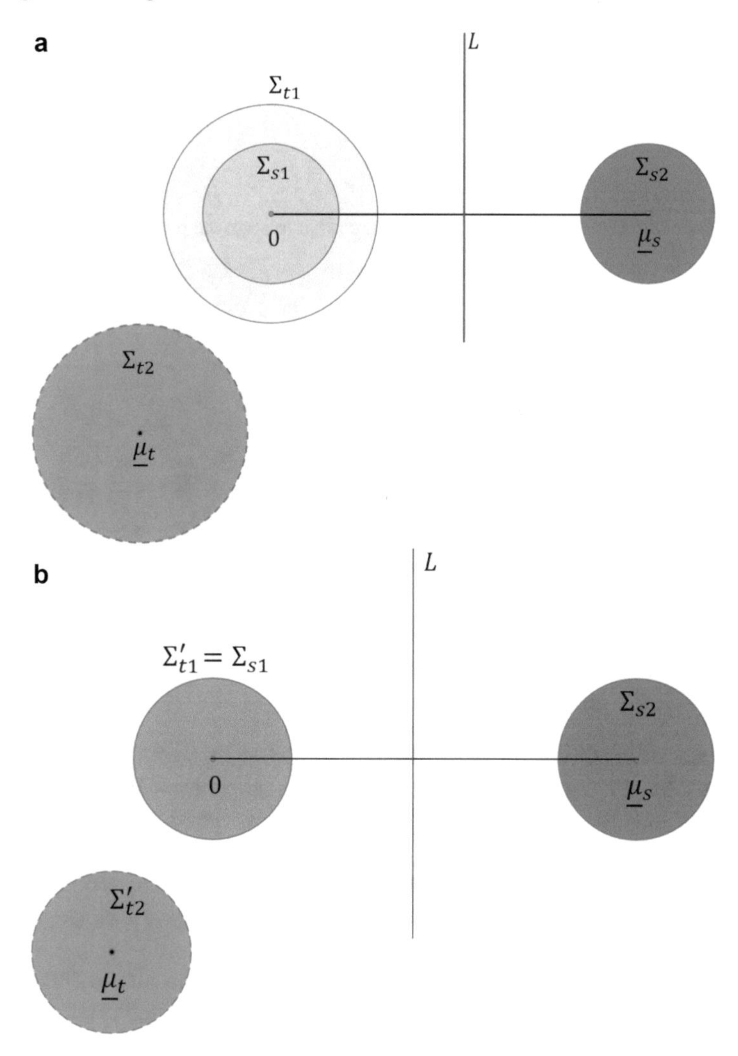

Fig. 8.6 Given *only* normal-condition data in the target domain: (**a**) shows the harmonising of means and (**b**) demonstrates the results of scaling the normal-condition clusters to coincide. The dotted border indicates a cluster for which there are no observations; the violet colour indicates that the two normal conditions are now completely harmonised

not known. In this case, a discriminator L trained on the source data will fail completely, so the problem can only be framed in terms of novelty. To completely harmonise the two normal conditions, one has the option to scale the source or target data, so that the covariances coincide. This situation is shown in Fig. 8.6b, where the scaling has been carried out on the target data; the new covariances for the target data are denoted Σ'_{t1} and Σ'_{t2}.

Scaling is not allowed within the group of affine transformations; however, it is allowed in the larger *conformal* group. The scaling has not helped in terms of discrimination; however, one would hope that most SHM problems would have more similar source–target pairs. Of course, novelty detection would still work. In the likely event that the normal-condition clusters were not spherical, the covariances could be aligned using rotation and scaling transformations on the target data. If desired, both the Class 1 and Class 2 supports could be mapped onto the unit circle. In fact, it should be clear that PTA is simply a generalisation of *standardisation*.

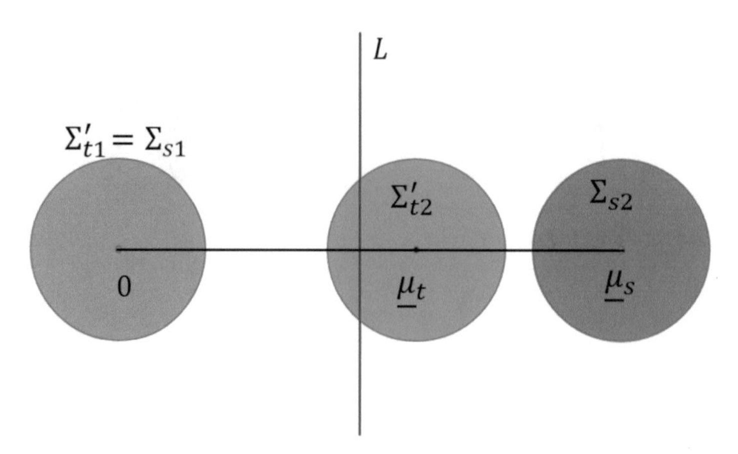

Fig. 8.7 In the presence of some target damage-state data, one can use a rotation to align the means

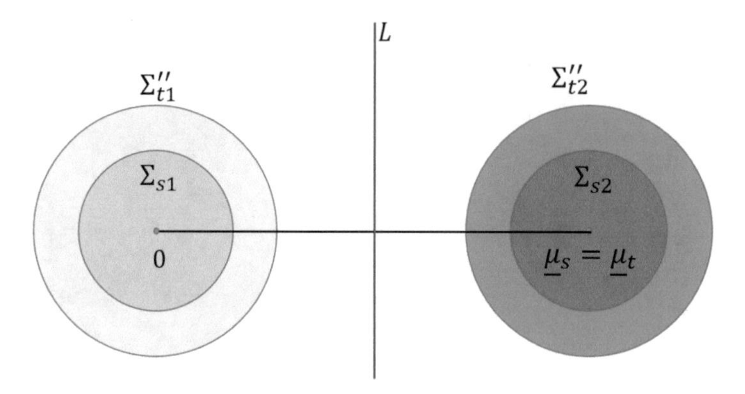

Fig. 8.8 Alternatively, with target damage-state data, one can scale to harmonise the means

Now, suppose a small amount of target damage-state data are available; enough to estimate $\underline{\mu}_t$; one can use the rotation degree-of-freedom on the target data to align the target clusters on the same axis as the source clusters, as shown in Fig. 8.7.

Here, a Fisher discriminant trained on the source problem will do some good but will make some mistakes. Of course, if enough target data are available to estimate accurate covariances, the line L could be moved in order to get perfect classification; however, in this situation one would not need transfer learning. In fact, another PTA strategy would be to use the scale degree-of-freedom on the target data to harmonise the means of the Class 2 clusters, as shown in Fig. 8.8. In this case, the Fisher discriminant on the source works perfectly.

In reality, covariances of the clusters will also be at play in PTA; however, in the simple illustration here, the availability of just enough target data to estimate $\underline{\mu}_{s1}$, $\underline{\mu}_{s2}$, $\underline{\mu}_{t1}$ and $\underline{\mu}_{t2}$ has enabled a solution to the problem without the need for domain adaptation. In fact, as discussed earlier, the term PTA is a bit of a misnomer, as some transfer has occurred above. In the more general case, one would hope that PTA would help further domain adaptation, rather than hinder it. This hypothesis is supported by recent work on more realistic examples [27]. In the general PBSHM framework, having chosen one or more source structures S_s^i for transfer, one can carry out any PTA operations likely to increase the possibility of positive transfer. Thus, the optimal strategy is likely to depend on metrics on the space of structures and feature spaces, and transformations on the latter.

8.6 Conclusions

The central point of this chapter is the observation that geometrical ideas—metrics and transformations—are both implicit and explicit in PBSHM. Apart from the fibre bundle providing an organising principle, the ideas discussed here are contributors to making *transfer* between structures effective. There are a number of principle points at which geometry is felt.

i. A metric on the space of structures can be used to select the 'closest' structure in a database (points in S) to the target structures of interest for TL, to minimise the possibility of negative transfer. One may also want to pick the k nearest neighbours and do multi-task learning.
ii. One may want to use a metric on the feature space as a proxy metric as defined earlier. If this is the case one ***should not*** carry out any scaling, standardisation or PTA on the data, as this will skew the metric assessment.
iii. Having chosen one or more source structures for transfer, one can carry out any PTA operations likely to increase the possibility of positive transfer. As seen earlier, these operations are based on the geometry of the feature spaces.
iv. Finally, the TL algorithm of interest may employ specific metrics itself; for example domain adaptation can be based on the MMD between $p(x)$ and $q(x)$ as embedded in an RKHS.

A significant issue that has been omitted here is that of confounding influences; this is a job for another day.

Acknowledgments This work has been supported by the UK Engineering and Physical Sciences Research Council (EPSRC), via the project *A New Partnership in Offshore Wind* (grant numbers EP/K003836/2 and EP/R004900/1). KW also gratefully acknowledges support from the EPSRC via grant no. EP/J016942/1, and EJC is grateful to EPSRC for support via grant no. EP/S001565/1.

References

1. Bull, L.A., Gardner, P.A., Gosliga, J., Rogers, T.J., Dervilis, N., Cross, E.J., Papatheou, E., Maguire, A.E., Campos, C., Worden, K.: Foundations of population-based SHM, part I: homogeneous populations and forms. Mech. Syst. Signal Process. **148**, 107141 (2021)
2. Gosliga, J., Gardner, P.A., Bull, L.A., Dervilis, N., Worden, K.: Foundations of population-based SHM, part II: heterogeneous populations – graphs, networks, and communities. Mech. Syst. Signal Process. **148**, 107144 (2021)
3. Gardner, P.A., Bull, L.A., Gosliga, J., Dervilis, N., Worden, K.: Foundations of population-based SHM, part III: heterogeneous populations – transfer and mapping. Mech. Syst. Signal Process. **149**, 107142 (2021)
4. Halmos, P.R.: Finite-Dimensional Vector Spaces. Springer, Berlin (1974)
5. Farrar, C.R., Worden, K.: Structural Health Monitoring: A Machine Learning Perspective. Wiley, New York (2011)
6. Worden, K., Manson, G.: Damage detection using outlier analysis. J. Sound Vib. **229**, 647–667 (2020)
7. Eguchi, T., Gilkey, P.B., Hanson, A.J.: Gravitation, gauge theories and differential geometry. Phys. Rep. **66**, 213–393 (1980)
8. Tsialiamanis, G., Mylonas, C., Chatzi, E., Dervilis, N., Wagg, D.J., Worden, K.: Foundations of population-based SHM, part IV: structures and feature spaces as geometry. Mech. Syst. Signal Process. **157**, 107692 (2021)
9. Berger, M.: A Panoramic View of Riemannian Geometry. Springer, Berlin (2002)
10. Caruna, R.: Multitask learning. Mach. Learn. **28**, 41–75 (1997)
11. Koch, I.: Enumerating all connected maximal common subgraphs in two graphs. Theor. Comput. Sci. **250**, 1–30 (2001)
12. Wills, P., Meyer, F.G.: Metrics for graph comparison: a practitioner's guide. Technical Report, arXiv:1904.07414v2 [stat.AP] (2020)
13. Gosliga, J., Hester, D., Worden, K., Bunce, A.: Population-based structural health monitoring for bridges. Mech. Syst. Signal Process. **173**, 108919 (2021)
14. Borgwardt, K.M., Ong, C.S., Schönauer, S., Vishwanathan, S.V.N., Smola, A.J., Kriegel, H.P.: Protein function prediction via graph kernels. Bioinformatics **21**, i47–i56 (2005)
15. Vishwanathan, S.V.N., Borgwardt, K.M., Schraudolph, N.N.: Fast computation of graph kernels. In: Advances in Neural Information Processing Systems, pp. 1449–1456 (2007)
16. Borgwardt, K.M., Kriegel, H.P.: Shortest-path kernels on graphs. In: Proceedings – IEEE International Conference on Data Mining, ICDM, pp. 74–81 (2007)
17. Borgwardt, K.M., Kriegel, H.P.: Scalable kernels for graphs with continuous attributes. In: Advances in Neural Information Processing Systems (2013)
18. Orsini, F., Frasconi, P., De Raedt, L.: Graph invariant kernels. In: IJCAI International Joint Conference on Artificial Intelligence, pp. 3756–3762 (2015)
19. Borgwardt, K.M., Gretton, A., Rasch, M.J., Kriegel, H.P., Schölkopf, B., Smola, A.J.: Integrating structured biological data by kernel maximum mean discrepancy. Bioinformatics **22**, 49–57 (2006)
20. Hartshorne, R.: Algebraic Geometry. Springer, Berlin (1971)
21. Silverman, B.W.: Density Estimation for Statistics and Data Analysis. Chapman and Hall, London (1986)
22. Bellman, R.E.. Dynamic Programming. Princeton University Press, Princeton (1957)
23. Gretton, A., Borgwardt, K.M., Rasch, M.J., Schölkopf, B., Smola, A.J.: A kernel two-sample test. J. Mach. Learn. Res. **13**, 723–773 (2012)
24. Pan, S.J., Yang, Q.: A survey on transfer learning. IEEE Trans. Knowl. Data Eng. **22**, 1345–1359 (2010)
25. Gardner, P.A., Liu, X., Worden, K.: On the application of domain adaptation in structural health monitoring. Mech. Syst. Signal Process. **138**, 1–24 (2020)
26. Rosenstein, M.T., Marx, Z., Kaelbling, L.P., Dietterich, T.G.: To transfer or not to transfer. In: NIPS 2005 Workshop on Transfer Learning, pp. 1–4 (2005)
27. Poole, J., Gardner, P.A., Dervilis, N., Bull, L.A., Worden, K.: On normalisation for domain adaptation in population-based structural health monitoring. In: Proceedings of 2021 International Workshop on SHM, Stanford, CA (2021)

Chapter 9
A Robust PCA-Based Framework for Long-Term Condition Monitoring of Civil Infrastructures

Mohsen Mousavi and Amir H. Gandomi

Abstract This chapter proposes an output-only method for condition monitoring of civil infrastructures through studying a couple of lowest structural natural frequency signals. The main challenge in this sort of problem is to mitigate the effect of the Environmental and Operational Variations (EOV) on the structural natural frequencies to avoid misinterpretation of these effects as damage. To this end, a robust Principal Component Analysis (PCA)-based approach is proposed that uses a couple of lowest structural natural frequency signals obtained from vibration data over a long period of time. First, the proposed method utilizes a truncated transformation matrix of the robust local PCA of a portion of the dataset corresponding to the healthy state of the structure to remove the EOV effects by mapping the dataset to a new space. The difference between the mapped signals and the original signals is deemed to minimize the effect of the EOV. As such, extracting the mapped data from the original data, termed error signals, will remove the EOV effects and can be further used for damage detection. To this end, the Mahalanobis distances of the errors in the test set from the distribution of the errors in the baseline data are used for condition monitoring through constructing a Hotelling (T^2) control chart. The proposed PCA-based method does not apply the covariance matrix and the mean vector of the entire dataset, but instead the Minimum Covariance Determinant (MCD) algorithm, in its fast mode (FastMCD), is employed to obtain a robust covariance matrix and mean vector of the dataset. It is shown through solving the benchmark problem of the Z24 bridge that the proposed method can effectively increase the accuracy of the damage detection compared with the case when the normal PCA is used.

Keywords SHM · Robust PCA · EOV · MCD · Hotelling charts

9.1 Introduction

Long-term condition monitoring of civil infrastructures can be classified based on the type of the employed data into two main categories: (1) output-only methods and (2) input–output methods [1]. The inputs are usually either the applied forces or ambient data such as temperature and humidity, whereas the outputs are structural responses either in time or frequency domain such as acceleration, natural frequencies, or mode shapes. The output-only methods have been given a great deal of attention lately, since they do not require multi-type sensors deployment on the structure under study to measure multi-type data. There have been different types of output-only methods proposed by researchers during the past decade. Some of the output-only methods for structural condition monitoring include: cointegration-based methods [2], principal component analysis-based methods [3], clustering methods [4], and outlier analysis-based methods [1].

Static and dynamic regression models complemented by a Principal Components Analysis (PCA) were employed for condition monitoring of the Infante D. Henrique Bridge using control charts [5]. There have been also some attempts to design and apply novel software to be used for long-term condition monitoring of structures [6]. Clustering, as a popular unsupervised technique, has been widely used for damage identification of structures. Some of the well-known clustering methods for structural damage detection include: k-means, k-medoids, Gaussian mixture model and fuzzy clustering [4, 7]. Likewise to the supervised methods, EOV can adversely affect the performance of unsupervised methods.

This chapter aims to develop a supervised condition monitoring algorithm that is robust to the EOV effects. The data used to this end is a couple of lowest natural frequencies of structure, identified over a long period of time. Knowing the EOV can compromise the effectiveness of condition monitoring methods, the main challenge is to develop a technique that can

M. Mousavi (✉) · A. H. Gandomi
Faculty of Engineering and Information Technology, University of Technology Sydney, Ultimo, NSW, Australia
e-mail: mohsen.mousavi@uts.edu.au

R. Madarshahian, F. Hemez (eds.), *Data Science in Engineering, Volume 9*, Conference Proceedings of the Society for Experimental Mechanics Series, https://doi.org/10.1007/978-3-031-04122-8_9

deal with the unwanted effects of EOV on the structural natural frequencies. This point is addressed through exploiting an advanced signal decomposition method, termed Variational Mode Decomposition (VMD) algorithm coupled with a robust PCA algorithm. To validate the performance capability of the proposed method, the condition monitoring problem of the Z24 bridge benchmark problem is studied. The superiority of the proposed method is demonstrated through comparison against the case where the normal PCA algorithm is used. The Novelties of this study can be thus summarised as follows:

1. A novel robust PCA-based structural condition monitoring framework is proposed. The proposed method outperforms a traditional version of itself.
2. The VMD algorithm is employed for removing seasonal patterns in structural natural frequency signals.
3. The FastMCD algorithm can produce slightly different values for the mean and covariance matrix of the dataset at each run. This will make the results of condition monitoring vary from one run of the computer program to another. Therefore, it is proposed to run the algorithm 100 times and the average and median of the results be taken for structural condition monitoring.
4. The proposed method was successfully tested on the Z24 bridge benchmark problem.

9.2 Background

The proposed method uses the Principal Component Analysis (PCA) of the structural natural frequency signals, identified over a long period of time, for condition monitoring of the structure. The proposed method is constructed based on a similar method presented in [5]. However, in this study, the robust scatter and location of the frequency signals are identified through employing a robust PCA algorithm and further used for damage identification. This strategy will be shown to be able to address the challenges imposed by the EOV effects on the structural condition monitoring problem. The proposed method is a baseline-dependent method, meaning that it requires baseline data from the healthy state of structure. Therefore, the proposed method has two main stages. The first stage considers constructing a baseline and obtaining a transformation matrix (to minimise the effect of the EOV on frequencies) based on data obtained from the healthy state of the structure. The obtained transformation matrix from the first stage is then used for conducting condition monitoring using data available from a secondary state of the structure.

It is known that the EOV effects mark mainly two different patterns on structural response data. These are: (1) a short-term seasonal pattern stemming from the short-term fluctuations of temperature or other EOV effects, and (2) a long-term pattern caused by the long-term fluctuations of such effects. Condition monitoring algorithms are adversely affected by the change of the variance of heteroscedastic data stemming from the seasonal effects, making the procedure of condition monitoring a complex task. Hence, the first stage of condition monitoring algorithms is usually aimed to remove complex seasonal patterns from structural response (natural frequencies in this chapter) [1, 2, 8, 9]. To this end, an advanced signal decomposition algorithm, termed Variational Mode Decomposition (VMD) [10], is used in this study. The VMD algorithm is generally used to decompose a signal $S(t)$ into its constructive oscillatory modes, termed Intrinsic Mode Functions (IMFs). Each IMF is narrow-band and can be thus characterised by its centre frequency ω. The algorithm of the VMD solves the following constraint variational optimisation problem:

$$\min_{\{u_k\}\ \&\ \{\omega_k\}} \sum_k \left\| \partial_t \left(\delta(t) + \frac{j}{\pi t} * u_k(t) \right) e^{-j\omega_k t} \right\|_2^2 ; \quad s.t. \ S(t) = \sum_k u_k(t) \tag{9.1}$$

in which $*$ is the convolution operator, j is the imaginary unit, u_k and ω_k are, respectively, the kth IMF and its centre frequency, $\delta(t)$ represents dirac distribution, and $||.||_2$ represents L^2-norm.

In this study, VMD is employed for two main reasons [1, 2, 9]: (1) to denoise the natural frequency signals, and (2) to remove the seasonal patterns in the signals. In order to solve the constraint optimisation problem of (9.1), a Lagrangian multiplier and a quadratic penalty term are added to the equation. This makes the VMD a parametric decomposition algorithm. The specified values of the VMD parameters, for the purpose of this study, are listed in Table 9.1. For further information about the ways of specifying VMD parameters, the readers are referred to [1, 2, 9].

The proposed method uses a robust PCA technique to obtain a transformation matrix out of a set of frequency signals corresponding to the healthy state of the structure under study. The obtained transformation matrix is then used in the second stage to remap the frequency signals pertaining to an unknown state of the structure to the original space, which can be further used to construct a damage-sensitive feature (DSF). To this end, the fast Minimum Covariance Determinant (FastMCD) algorithm [11] is employed to derive the robust covariance matrix and mean vector pertaining to the matrix of all observations.

Table 9.1 Specified values for the VMD parameters

Parameters	Description	Specified values
K	Number of IMFs	2
α	Denoising factor	100
τ	Time interval	0
ϵ	Convergence threshold	10^{-5}
$init$	Centre frequency initialiser	0
DC	Boolean parameter	0

Figure 9.1 shows the steps of the proposed condition monitoring framework. It is assumed that some data (a couple of lowest natural frequencies) from the healthy state of the structure are available as baseline in the first step. As such, the first step is outlined as follows:

1. Identify a couple of lowest natural frequencies of the healthy structure from the vibration signals at some instants over a long period of time and set them in the matrix $\mathbf{\Omega}_1$. To this end, the stochastic subspace identification method is usually employed [12].
2. Employ the VMD algorithm to denoise the signals and remove the seasonal patterns from each of the columns of $\mathbf{\Omega}_1$.
3. Employ the FastMCD algorithm to obtain the robust covariance matrix of $\mathbf{\Omega}_1$ as Σ_1.
4. Obtain the Singular Value Decomposition (SVD) of Σ_1 as $\mathbf{\Sigma}_1 = \mathbf{U}_1\mathbf{S}_1\mathbf{U}_1^{'}$, where $\mathbf{S}$ is the diagonal matrix of eigenvalues, $\mathbf{U}$ is the matrix of corresponding eigenvectors, and $'$ denotes the transpose of a matrix.
5. Obtain a truncated form of the transformation matrix $\mathbf{U}_1$, shown as $\hat{\mathbf{U}}_1$, through eliminating the columns corresponding to the principal components of higher orders. Note that in this chapter, the first two columns corresponding to the PC_1 and PC_2 are retained.

The second step of the proposed method is outlined as follows:

1. Identify the same number of lowest natural frequencies of the structure, likewise to the first stage, from the recorded vibration data and set them in the matrix $\mathbf{\Omega}_2$.
2. Employ the VMD algorithm to denoise and remove the seasonal patterns of the columns of $\mathbf{\Omega}_2$.
3. Employ the truncated $\hat{\mathbf{U}}_1$, obtained from the first stage, to map the natural frequency signals of the second stage as follows:

$$\hat{\mathbf{\Omega}}_2 = \mathbf{\Omega}_2\, \hat{\mathbf{U}}_1\, \hat{\mathbf{U}}_1^{'} \tag{9.2}$$

Likewise, the same procedure is followed for the first stage as follows:

$$\hat{\mathbf{\Omega}}_1 = \mathbf{\Omega}_1\, \hat{\mathbf{U}}_1\, \hat{\mathbf{U}}_1^{'} \tag{9.3}$$

4. Consider the error associated with this mapping as a damage-sensitive feature (DSF), as follows:

$$\Delta\mathbf{\Omega}_2 = |\hat{\mathbf{\Omega}}_2 - \mathbf{\Omega}_2| \tag{9.4}$$

Likewise,

$$\Delta\mathbf{\Omega}_1 = |\hat{\mathbf{\Omega}}_1 - \mathbf{\Omega}_1| \tag{9.5}$$

where $|.|$ is the absolute value operator.

Finally, a new signal needs to be constructed out of the mapped frequency matrix for damage identification. This is done through using the concept of Shewhart (T^2) charts as follows:

The Phase II Shewhart (T^2) chart is proposed to be employed for obtaining a signal out of the mapped frequency signals, which is suitable for condition monitoring of structures [13, 14]. The T^2 chart, constructed for each individual observation in the second stage (vector $\Delta\mathbf{\Omega}_2$), is the second power of the Mahalanobis distance of that observation from the average point of the dataset corresponding to the first stage (vector $\Delta\mathbf{\Omega}_1$) as follows:

$$T^2 = (\Delta\mathbf{\Omega}_2 - \boldsymbol{\mu}_1)^{'}\ \mathbf{\Sigma}_1^{-1}\ (\Delta\mathbf{\Omega}_2 - \boldsymbol{\mu}_1) \tag{9.6}$$

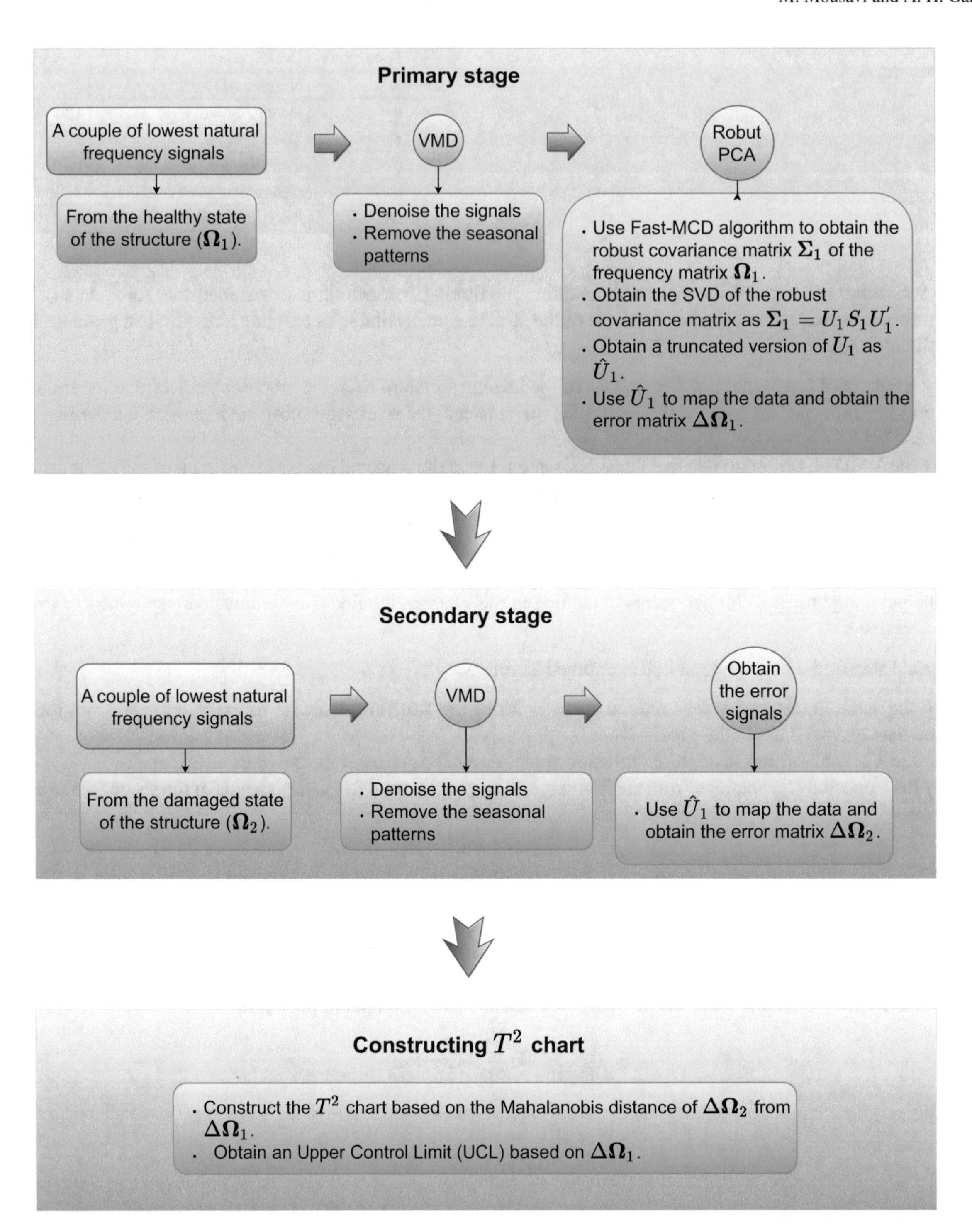

Fig. 9.1 The scheme of the proposed method

where $\boldsymbol{\mu}_1$ and $\mathbf{\Sigma}_1$ are, respectively, the mean vector and covariance matrix of $\Delta\mathbf{\Omega}_1$ obtained through the FastMCD algorithm. An Upper Control Limit (UCL) for the constructed chart can be obtained based on information corresponding to the first state of the structure as follows:

$$\mathrm{UCL} = \frac{p(m+1)(m-1)}{m^2 - mp} \times F_{\alpha;\, p,m-p} \tag{9.7}$$

where p is the number of variables-identified natural frequencies at each time instant, $F_{\alpha;\ p,m-p}$ represents the value of the F-distribution with p and $m-p$ degrees of freedom, and α is the confidence level. To get a 95 % confidence for the UCL, α was set to 0.05, and m is the number of observations in the first stage, which corresponds to the healthy state of the structure.

9.3 Analysis

In this section, the problem of condition monitoring of the Z24 bridge is studied. To this end, the four lowest natural frequency signals of the structure are analysed using the proposed method for condition monitoring of the structure. The benchmark problem of the Z24 bridge has been used in many applications such as those aimed at removing the influence of the EOV effects on structural modal data [15–17]. The corresponding data of the structural natural frequencies can be found upon request from "https://bwk.kuleuven.be/bwm/z24".

First, the results of the application of the normal PCA are presented in Fig. 9.2a. As is evident from the figure, the results of the condition monitoring is not satisfactory, as the numbers of false positive/negative cases are plentiful. Next, the robust PCA was employed in the proposed condition monitoring framework. To this end, the entire dataset was divided into two parts, where the first 50% portion of the dataset was used for obtaining the truncated matrix $\hat{\mathbf{U}}_1$, as discussed in the previous section. Note that the first 50% dataset thus corresponds to the healthy state of the structure. The remaining 50% portion of the dataset thus include information about damage. First, the effect of the EOV on the natural frequency signals was eliminated using the proposed PCA-based approach. Finally, Eq. 9.6 was used to obtain the T^2 chart of the second portion of the dataset, for which the UCL was obtained via Eq. 9.7. Note that the FastMCD algorithm can produce marginally different results in different attempts, which can affect the final damage identification results. Here, a typical obtained result is shown in Fig. 9.2b, though later on we propose a strategy to deal with this problem. As is evident from the results, the proposed method is far more successful in condition monitoring of the Z24 bridge. In order to quantify the results, the F_1 score was employed, which is defined as follows:

$$F_1 = \frac{tp}{tp + \frac{1}{2}(fn + fp)} \tag{9.8}$$

where tp, fn, and fp denote, respectively, the number of true positive, false negative, and false positive cases in the obtained results. The F_1 score was, respectively, calculated for the normal and robust PCA-based methods' results as 0.0996 and 0.9257, indicating the far better performance of the proposed condition monitoring framework. Note that in both cases VMD was employed for removing the seasonal patterns from the dataset. Reiterated, the only downside of the proposed strategy is that the results can vary in different runs of the FastMCD algorithm, making the final decision of the condition monitoring

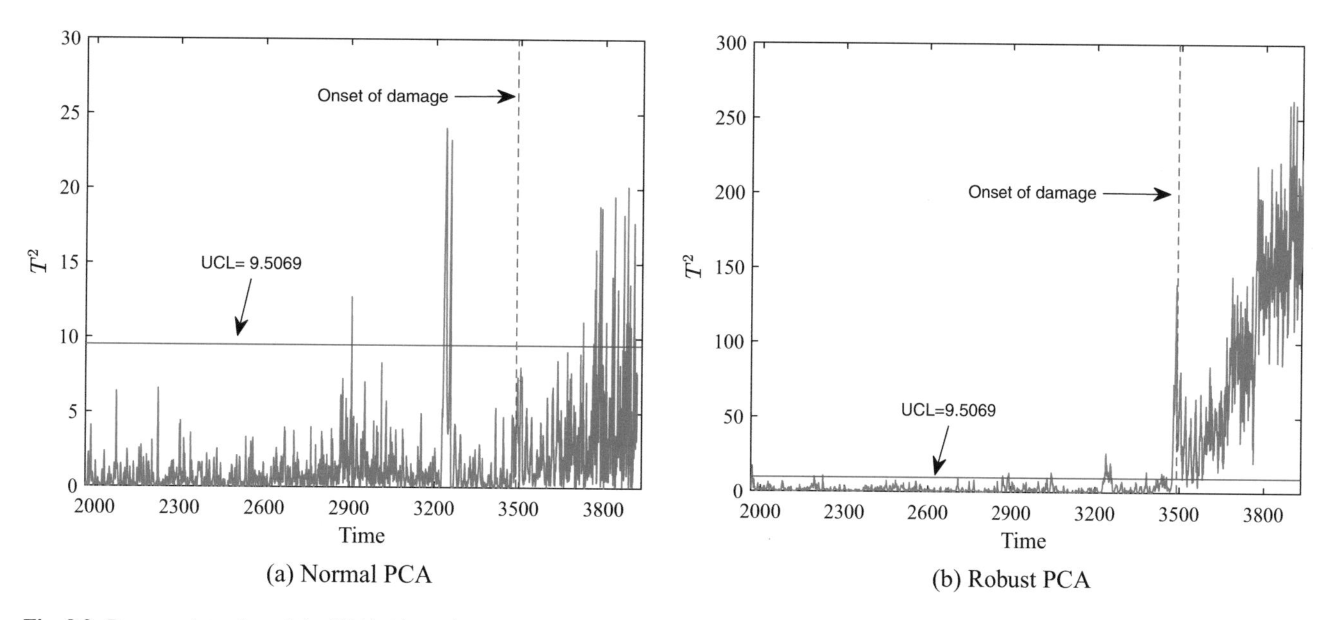

Fig. 9.2 Damage detection of the Z24 bridge using (**a**) normal and (**b**) robust PCA

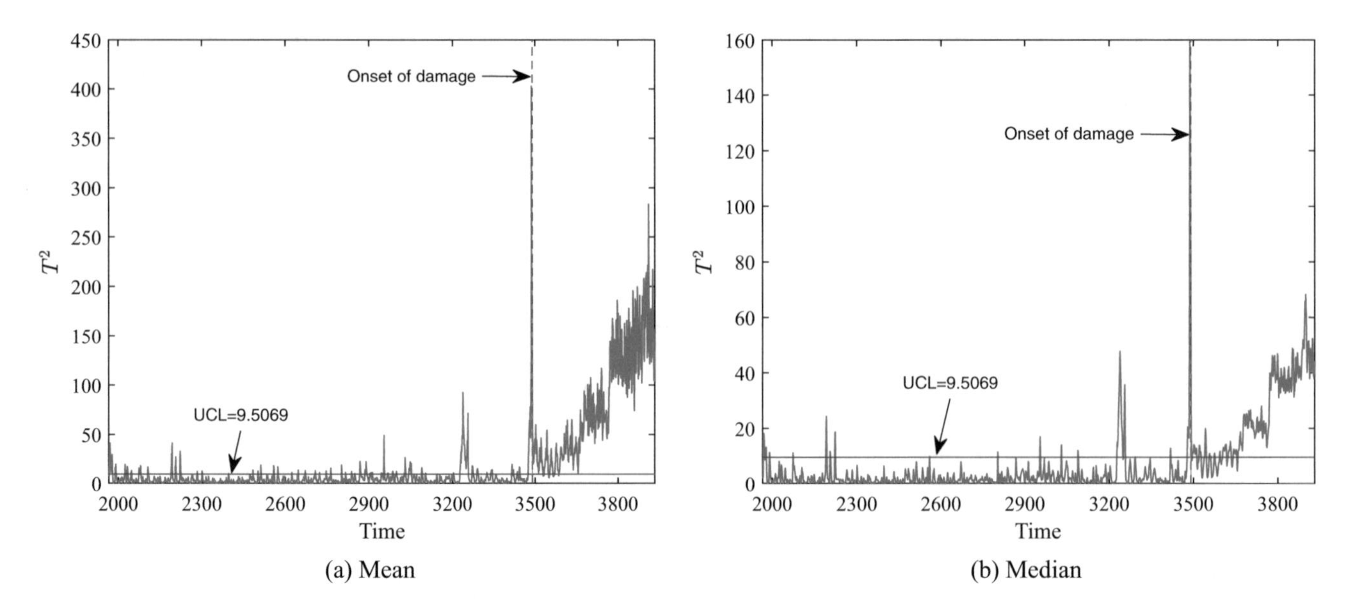

Fig. 9.3 Damage detection of the Z24 bridge using the proposed robust PCA method. The algorithm was ran 100 times and the mean (**a**) and median (**b**) of the final results were taken

challenging. To confront this problem, it is recommended that the algorithm be run multiple times and the average value as well as the median of the final results be taken. Here, the algorithm was run 100 times. The mean and the median of the final results are presented in Fig. 9.3.

Next, the F_1 score for the mean and median of the results was calculated. This value was obtained, respectively, for the mean and median of the results as 0.7915 and 0.8069. This truly again demonstrates the superior performance of the proposed strategy compared with the case when the normal PCA was employed (with $F_1 = 0.0996$) for condition monitoring of the Z24 bridge.

Note that although some false positive/negative cases can be identified in the final results of the both Fig. 9.3a, b, one can ignore the short-term violations of the threshold as they cannot be referred to damaged/undamaged cases when the results are not persistent. This is mainly due to the fact that damage is believed to have a stationary effect on the results and, therefore, its presence cannot appear in the final results as a short-term violation of the threshold [5] (likewise to its absence). As such, one may devise a better strategy for quantifying the performance of the proposed condition monitoring strategy than the F_1 score. Since all such fp and fn cases were ruled in calculation of the F_1 score, it is actually a worst case evaluation criteria.

9.4 Conclusion

A robust PCA-based condition monitoring method for identification of anomalous observations of the natural frequencies that can be referred to damage has been proposed. The proposed method utilizes the FastMCD algorithm to obtain the robust covariance matrix of the healthy state of structure to obtain a transformation matrix to be further used for transforming data to a new space. Once the data from the secondary state of the structure was transformed into the new space, the error associated with this transformation was considered as a DSF. The Mahalanobis distances of the errors pertaining to the secondary state of the structure from the primary state (healthy sate) of the structure (T^2 chart) were monitored for condition monitoring of the structure. Since the results of the FastMCD algorithm can vary from one run of the algorithm to another, the FastMCD algorithm was run 100 times and the average and median of the results were considered for the condition monitoring of the structure. The results demonstrate the superior performance of the proposed method compared with the case when a normal PCA was employed in the calculations. The authors aim to improve the performance of the proposed condition monitoring strategy through the application of other methods for robust identification of the scatter and location of dataset.

Acknowledgment The KU Leuven Structural Mechanics Section is acknowledged as the source of the data for Z24 bridge.

References

1. Mousavi, M., Gandomi, A.H.: Structural health monitoring under environmental and operational variations using MCD prediction error. J. Sound Vib. **512**, 116370 (2021)
2. Mousavi, M., Gandomi, A.H.: Prediction error of Johansen cointegration residuals for structural health monitoring. Mech. Syst. Signal Process. **160**, 107847 (2021)
3. Roberts, C., Garcia, D., Tcherniak, D.: A comparative study on data manipulation in PCA-based structural health monitoring systems for removing environmental and operational variations. In: Proceedings of the 13th International Conference on Damage Assessment of Structures, pp. 182–198. Springer, Berlin (2020)
4. Sarmadi, H., Entezami, A., Salar, M., De Michele, C.: Bridge health monitoring in environmental variability by new clustering and threshold estimation methods. J. Civil Struct. Health Monit. **11**, 1–16 (2021)
5. Magalhães, F., Cunha, Á., Caetano, E.: Vibration based structural health monitoring of an arch bridge: from automated OMA to damage detection. Mech. Syst. Signal Process. **28**, 212–228 (2012)
6. García-Macías, E., Ubertini, F.: MOVA/MOSS: two integrated software solutions for comprehensive structural health monitoring of structures. Mech. Syst. Signal Process. **143**, 106830 (2020)
7. de Almeida Cardoso, R., Cury, A., Barbosa, F.: Automated real-time damage detection strategy using raw dynamic measurements. Eng. Struct. **196**, 109364 (2019)
8. Shi, H., Worden, K., Cross, E.J.: A cointegration approach for heteroscedastic data based on a time series decomposition: an application to structural health monitoring. Mech. Syst. Signal Process. **120**, 16–31 (2019)
9. Mousavi, M., Gandomi, A.H.: Deep learning for structural health monitoring under environmental and operational variations. In: Nondestructive Characterization and Monitoring of Advanced Materials, Aerospace, Civil Infrastructure, and Transportation XV, vol. 11592, p. 115920H. International Society for Optics and Photonics (2021)
10. Dragomiretskiy, K., Zosso, D.: Variational mode decomposition. IEEE Trans. Signal Process. **62**(3), 531–544 (2014)
11. Hubert, M., Debruyne, M., Rousseeuw, P.J.: Minimum covariance determinant and extensions. Wiley Interdiscip. Rev. Comput. Stat. **10**(3), e1421 (2018)
12. Cancelli, A., Laflamme, S., Alipour, A., Sritharan, S., Ubertini, F.: Vibration-based damage localization and quantification in a pretensioned concrete girder using stochastic subspace identification and particle swarm model updating. Struct. Health Monit. **19**(2), 587–605 (2020)
13. Thomas, R.: Statistical Methods for Quality Improvement, 2nd edn. Wiley, New York (2000)
14. Montgomery, D.C.: Introduction to Statistical Quality Control, Chap. 10, 4th edn. Wiley, New York (2001)
15. Peeters, B., De Roeck, G.: One-year monitoring of the z24-bridge: environmental effects versus damage events. Earthq. Eng. Struct. Dyn. **30**(2), 149–171 (2001)
16. Reynders, E., Wursten, G., De Roeck, G.: Output-only structural health monitoring in changing environmental conditions by means of nonlinear system identification. Struct. Health Monit. **13**(1), 82–93 (2014)
17. Langone, R., Reynders, E., Mehrkanoon, S., Suykens, J.A.: Automated structural health monitoring based on adaptive kernel spectral clustering. Mech. Syst. Signal Process. **90**, 64–78 (2017)

Chapter 10
Data-Driven Structural Identification for Turbomachinery Blisks

Sean T. Kelly and Bogdan I. Epureanu

Abstract Blisks are commonly used within compressors of modern turbomachinery and are nominally cyclic symmetric structures (i.e., tuned) made from a single piece of material with uniform sector-to-sector material properties and geometry. However, due to manufacturing tolerances, blisks contain sector-to-sector perturbations in material properties and geometry known as mistuning, which can result in increased response amplitudes due to energy localization, causing greater stresses and risk of high cycle fatigue failure. As such, identifying structural properties such as mistuning of as-manufactured blisks is crucial for blisk design and accurately predicting blisk dynamics in operation. Previously, we have presented a data-driven approach using a single feed-forward neural network designed for identifying mistuning of blisks, where mistuning is defined as perturbations in the blade-alone Young's modulus which vary randomly from blade to blade. To identify the mistuning within each sector, this approach only uses physical-response data from an individual sector as well as forcing information from traveling-wave excitations. Unlike most previous approaches for blisk mistuning identification, no modal information is used from either individual sectors or for the entire blisk nor blade isolation and/or detuning via added masses or damping pads. The general techniques presented in the past are augmented with additional studies exploring separate cases of response and forcing data inputs, and a discussion of the applicability of this data-driven approach is provided. These studies include further analysis of forcing data input requirements for the neural network and using response data targeting different modes of the underlying tuned system with and without including resonance peaks and significant measurement noise.

Keywords Data-driven · Turbomachinery · Blisks · Mistuning · Neural networks

10.1 Introduction

Nominally, turbomachinery bladed disks, or blisks, or cyclic-symmetric structures manufactured as a single piece. However, due to small inherent geometric deviations due to manufacturing tolerances and variabilities in material properties, called mistuning, the cyclic-symmetry of blisks is destroyed [1]. These deviations can lead to significant energy localization during operation when subject to traveling-wave excitation, leading to increased stresses from increased vibration amplitudes and increased risk of high cycle fatigue failure [2]. To predict mistuned blisk dynamics, several reduced-order models (ROMs) have been developed [3, 4]. However, to adapt these ROMs for a specific as-manufactured blisk, identification (ID) of mistuning is required. For bladed disks with inserted blades, individual blades can be tested separately in isolated conditions to identify these deviations. However, because blisks are manufactured as a single component, these techniques cannot be used, motivating the development of mistuning ID methods for blisks specifically [5–7]. These methods need to account for significant coupling between sectors due to inherently low damping resulting from a lack of contact interfaces, as well as geometric constraints. These previous methods often require isolating individual blades using added masses (i.e., mass detuning) or damping pads, as well as either blade-alone or system-level modal response information. However, the low internal damping of blisks makes sector isolation difficult to achieve, resulting in residual coupling effects that can comprise the accuracy of the ID. Further, modal response information can be difficult to obtain without significant noise, especially in regions of high modal density as are often present in blisks [2, 8].

To avoid these potential difficulties, previously the authors proposed a data-driven approach using a single feed-forward neural network (FFNN) for mistuning ID, which unlike previous approaches does not require any modal information nor

S. T. Kelly (✉) · B. I. Epureanu
Department of Mechanical Engineering, University of Michigan, Ann Arbor, MI, USA
e-mail: seantk@umich.edu; epureanu@umich.edu

R. Madarshahian, F. Hemez (eds.), *Data Science in Engineering, Volume 9*, Conference Proceedings of the Society for Experimental Mechanics Series, https://doi.org/10.1007/978-3-031-04122-8_10

any blade isolation techniques [9]. This data-driven approach is sector level, whereby the mistuning is identified with only physical response and forcing information from within a single sector. This data is generated from a traveling-wave excitation like that seen during operation [2]. To validate this approach previously, we considered an excitation frequency range containing only blade-dominated response information from an isolated mode family. To augment the previous work, we now consider responses from higher order modes, which are not isolated, targeting modes with response information that can be either disk or blade-dominated. We also explore the effects of varying the forcing reference relative the responses as inputs to this network, as well as prediction with and without including resonant peaks. These studies are validated using computational surrogate data from a Finite Element (FE) model with up to 10% relative and 5% absolute measurement noise in the data.

10.2 Background and Methodology

As has been done for many previous studies, small stiffness mistuning parametrized by deviations from a nominal blade-alone Young's modulus E_o is considered. This mistuning δ^i for blade i is calculated as $\delta^i = \sigma r^i$, where σ is the mistuning magnitude and r^i is a randomly generated value from a Gaussian distribution with mean 0 and standard deviation 1. This mistuning is applied to give the mistuned blade-alone Young's modulus for blade i as $E_i = (1 + \delta^i)E_o$ as shown in Fig. 10.1.

For this data-driven mistuning ID approach, we present only a general discussion of the overall methodology. However, previous studies and a more comprehensive overview and discussion of this approach are presented in [9]. For this data-driven approach, we consider a single sector-level FFNN which serves to map a set of inputs to a set of outputs. For a general sector i, the inputs consist of physical responses along the disk to account for coupling from neighboring sectors, as well as the traveling-wave forcing frequency ω_j and forcing phase $\angle f_i$. The output is the mistuning value δ^i. The responses considered for all work presented are shown in Fig. 10.1, where only out-of-plane (i.e., axial) responses are considered at these selected points. Only response data from single points along the high and low interfaces and the blade tip are used, which is the minimum number of points required to capture both coupling effects as well as the blade dynamics. For this approach to achieve good ID accuracy, only response data from excitation frequencies which often (across all sectors and training mistuning patterns) result in large responses with low noise-to-signal ratios are considered. Additionally, if significant measurement noise is present, responses need to be conditioned using multiple trials. The full systematic procedures for frequency selection and response conditioning are provided in [9]. Lastly, because we consider a range of excitation frequencies but the mistuning δ^i is assumed constant, the output δ^i from the FFNN for a sector i across all excitation

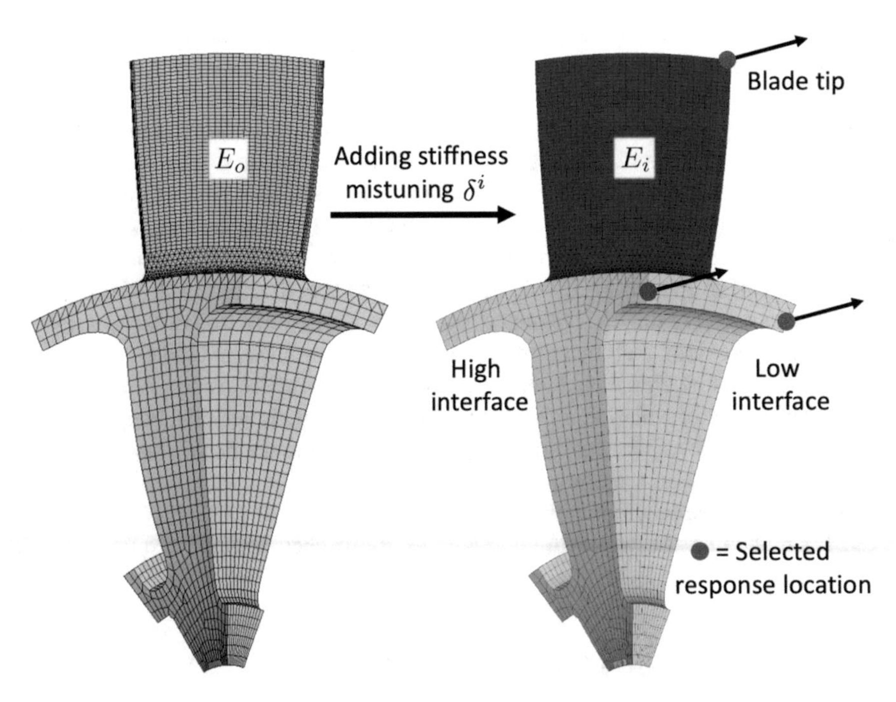

Fig. 10.1 Applied blade-alone stiffness mistuning and selected response locations

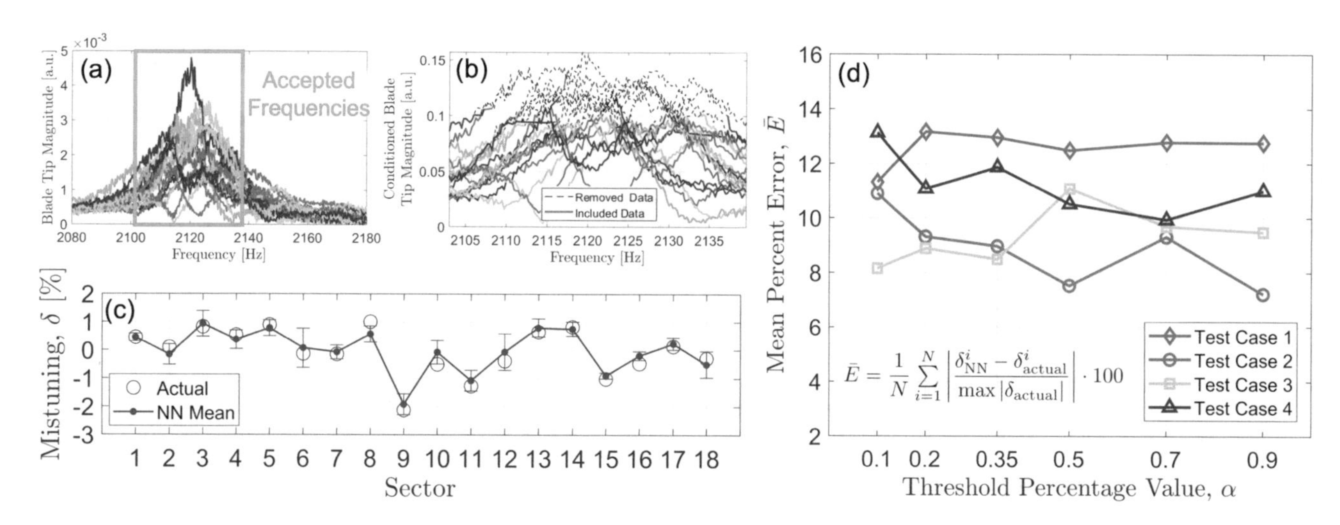

Fig. 10.2 (**a**) Original noisy and (**b**) conditioned blade tip responses (resonance peaks removed in (**b**)); (**c**) example ID using responses in (**b**); (**d**) example mean ID percent errors vs. frequency threshold percentage for removed resonance peaks

frequencies is averaged to give the final identified mistuning. This averaging is what allows this approach to not require the resonance peak for a blade unlike many previous methods while also providing confidence bounds for each identified δ^i using the standard deviations from the averaging process.

10.3 Results

The FE blisk model considered contains 18 sectors, and, unlike previous work in [9], we separately consider engine-order (EO) 2 and 4 excitations in a frequency range containing natural frequencies of modes across three higher mode families (see Fig. 10.2a for example noisy blade tip responses with frequencies accepted for ID indicated). This frequency range contains both torsional and bending modes, with EO2 excitation giving disk-dominated responses and EO4 blade-dominated responses. For these excitations, we first consider networks trained with a forcing phase reference shifted relative to those used for response generation. Next, we show mistuning ID results with and without using resonance peak response data (see Fig. 10.2b for example conditioned blade tip response data with all resonance peaks removed) from the conditioned responses. Considering response data with 10% relative and 5% absolute measurement noise and three trials for response conditioning, for all studies presented, we show mistuning ID can be performed accurately (see Fig. 10.2c for ID of example test case mistuning pattern). We additionally show low mean mistuning ID errors for four test cases varying the frequency threshold percentage α (see [9] for more details and Fig. 10.2d for example errors) with and without resonance peaks removed from the conditioned response data.

10.4 Conclusions

Studies using a data-driven sector-level approach for mistuning ID using a FFNN were presented. These include using noisy single-EO response data that can be either disk or blade-dominated for an excitation frequency range containing multiple higher mode families. In this frequency range, we showed that mistuning ID can be performed accurately with this approach while: (1) training with a shifted forcing phase reference relative to the response data and (2) obtaining mistuning ID values with or without including resonance peaks from the conditioned response data.

Acknowledgments The authors would like to gratefully acknowledge the generous support offered by the GUIde 6 Consortium toward their investigation and research in this topic and ANSYS Inc. for providing software licenses used for full system simulations.

References

1. Castanier, M.P., Pierre, C.: Modeling and analysis of mistuned bladed disk vibration: current status and emerging directions. J. Propuls. Power. **22**(2), 384–396 (2006). https://doi.org/10.2514/1.16345
2. Jones, K.W., Cross, C.J.: Traveling wave excitation system for bladed disks. J. Propuls. Power. **19**(1), 135–141 (2003). https://doi.org/10.2514/2.6089
3. Lim, S.-H., Bladh, R., Castanier, M.P., Pierre, C.: Compact, generalized component mode mistuning representation for modeling bladed disk vibration. AIAA J. **45**(9), 2285–2298 (2007). https://doi.org/10.2514/1.13172
4. Madden, A., Epureanu, B.I., Filippi, S.: Reduced-order modeling approach for blisks with large mass, stiffness, and geometric mistuning. AIAA J. **50**(2), 366–374 (2012). https://doi.org/10.2514/1.J051140
5. Feiner, D.M., Griffin, J.H.: Mistuning identification of bladed disks using a fundamental mistuning model—Part I: Theory. J. Turbomach. **126**(1), 150–158 (2004). https://doi.org/10.1115/1.1643913
6. Judge, J.A., Pierre, C., Ceccio, S.L.: Experimental mistuning identification in bladed disks using a component-mode-based reduced-order model. AIAA J. **47**(5), 1277–1287 (2009). https://doi.org/10.2514/1.41214
7. Figaschewsky, F., Kühhorn, A., Beirow, B., Giersch, T., Schrape, S., Nipkau, J.: An inverse approach to identify tuned aerodynamic damping, system frequencies and mistuning: Part 3—Application to engine data, presented at the ASME Turbo Expo 2019: turbomachinery technical conference and exposition, 2019. https://doi.org/10.1115/GT2019-91337
8. Yumer, M.E., Cigeroglu, E., Özgüven, H.N.: Mistuning identification of integrally bladed disks with cascaded optimization and neural networks. J. Turbomach. **135**(3), 031008 (2013). https://doi.org/10.1115/1.4006667
9. Kelly, S.T., Lupini, A., Epureanu, B.I.: Data-Driven Approach for Identifying Mistuning in as-Manufactured Blisks, Presented at the ASME Turbo Expo 2021: Turbomachinery Technical Conference and Exposition, 2021. https://doi.org/10.1115/GT2021-59887

Chapter 11
Classification of Rail Irregularities from Axle Box Accelerations Using Random Forests and Convolutional Neural Networks

Cyprien Hoelzl, Lucian Ancu, Henri Grossmann, Davide Ferrari, Vasilis Dertimanis, and Eleni Chatzi

Abstract The continuously increasing demand for mobility results in increased loading of the Swiss railway network, which is further associated with higher wear and deterioration of the rail infrastructure. Safety relevant surface defects on railway tracks, such as squats, have acted as an important driver of rail replacements in Europe. The early detection of such defects can support the planning of appropriate maintenance measures, such as grinding, which prolong the remaining life of the rails. On-board monitoring has redefined the paradigm of railway infrastructure monitoring, via use of in-service vehicles as mobile sensing systems. Such vehicles are equipped with sensors, e.g. axle box accelerometers in order to continuously collect information on the track and vehicle condition, and support the monitoring of railway assets and infrastructure. Acceleration-based monitoring has been shown to bear tremendous potential for offering temporally and spatially dense diagnostics of railway infrastructure. While the potential of such a monitoring scheme has been proven, the generalization has been limited due to the small sample sizes in existing studies.

We propose a methodology to recognize and classify between the most common rail irregularities, namely surface defects, insulated joints and welds, by exclusively relying on the availability of on-board acceleration measurements. We combine labeled information, stemming from rail-head image-based detection, with acceleration measurements. Two classification approaches are compared in this work. The first methodology exploits Convolutional Neural Networks (CNNs) that are applied to the Fourier coefficients, which are computed from acceleration time-series data. The second methodology relies on a more classical machine learning approach, applied on features that are extracted from the acceleration time series, which are then classified using Random Forests. Finally, the uncertainty of the acceleration metrics and of the ground-truth labels is analyzed, motivating the application of acceleration-based detection for improvement of rail condition monitoring. The resulting classifiers can be deployed on regular passenger trains for enabling the continuous and automated monitoring of the rail condition.

Keywords Rail and track dynamics · Machine learning · Big data · Data-driven diagnostics · Condition monitoring · Damage assessment

11.1 Introduction

Frequent monitoring to assess the condition of the rails is indispensable for applying predictive maintenance. Most railway tracks consist of continuously welded rails that are discretely supported onto sleepers. In some cases, instead of *welds*, the rails are connected using *insulated joints*, which consist of two electrically insulated rails that are bolted together with steel plates. The appearance of a fault usually initiates with small material imperfections that under accumulated load grow to more severe faults. *Surface defects* are a broad category of defects that have many origins. They can be caused by ballast on the rail surface, lost goods, or damaged wheels leading to indentations on the rail. It has been observed that some surface defects grow into a squat. *Squats* are defined by the International Union of Railways as a "widening and localized depression of the rail/wheel contact band, accompanied by a dark spot containing cracks with a circular arc or V shape" [1]. *Cracks* to

C. Hoelzl (✉) · H. Grossmann · D. Ferrari · V. Dertimanis · E. Chatzi
Institute of Structural Engineering, Department of Civil, Environmental and Geomatic Engineering, ETH Zürich, Zürich, Switzerland
e-mail: hoelzl@ibk.baug.ethz.ch; henri@student.ethz.ch; dferrari@student.ethz.ch; v.derti@ibk.baug.ethz.ch; chatzi@ibk.baug.ethz.ch

L. Ancu
Measurement and Diagnostics, Swiss Federal Railways (SBB), Bern, Switzerland
e-mail: lucian_stefan.ancu@sbb.ch

R. Madarshahian, F. Hemez (eds.), *Data Science in Engineering, Volume 9*, Conference Proceedings of the Society for Experimental Mechanics Series, https://doi.org/10.1007/978-3-031-04122-8_11

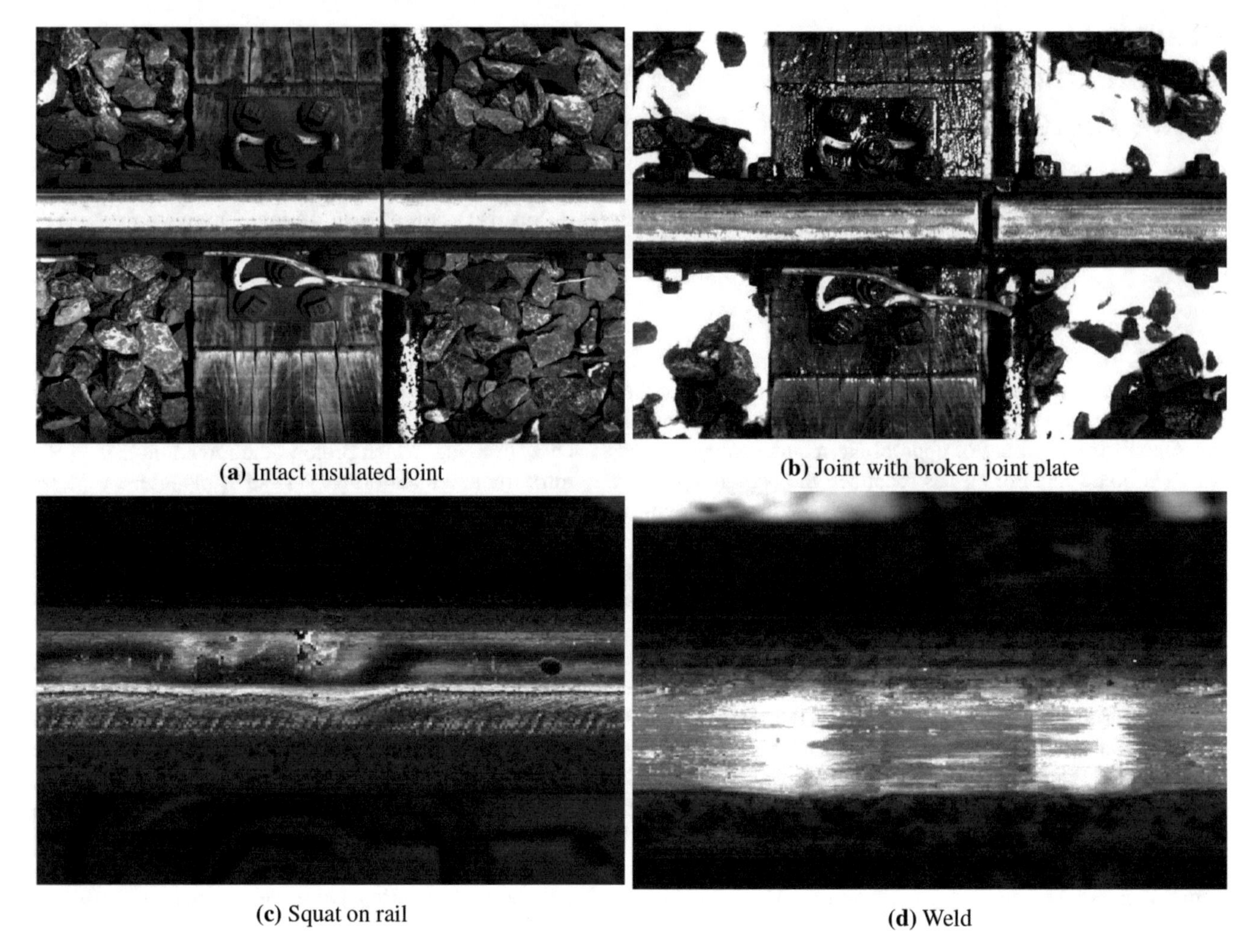

(a) Intact insulated joint
(b) Joint with broken joint plate
(c) Squat on rail
(d) Weld

Fig. 11.1 Characteristic defect and component classes (images extracted from a high speed camera, mounted on a diagnostic vehicle of the SBB). (**a**) Intact insulated joint. (**b**) Joint with broken joint plate. (**c**) Squat on rail. (**d**) Weld

the head, web, foot, weld, and joint plates are events that can lead to a broken rail. This type of flaw is a more rare occurrence that is further elaborated in the UIC Code 712 [1] (Fig. 11.1).

Diagnostic vehicles and their low cost counterpart—OBM vehicles—have been introduced to measure and collect data relating to the track state. Diagnostic vehicles are equipped with sensitive and highly accurate measurement systems but require specially organized measurement rides and periodic system maintenance. OBM vehicles on the other hand are in-service vehicles equipped with simple sensors such as accelerometers, to continuously gather data related to track and vehicle condition.

Two approaches are generally followed when dealing with identifying and classifying accelerations time series: parametric and non-parametric methods. Parametric methods, such as the Kalman filter, can be used to reconstruct the longitudinal level profile [2]. The longitudinal level is obtained via double integration of axle box accelerations (ABA) [3]. Non-parametric methods include time frequency domain analysis methods such as Fast Fourier Transform (FFT), Short Time Fourier Transform (STFT), Discrete Wavelet Transform (DWT), or Continuous Wavelet Transform (CWT). Molodova [4] used the wavelet coefficients from the CWT to classify squats and welds from axle box acceleration signals. This generalization of a classifier requires a dataset of sufficient size containing various types of track, welds, defects, and operational conditions.

The methodology we propose to recognize and classify between the most common rail irregularities, namely surface defects, insulated joints, and welds exclusively relies on axle box acceleration measurements from the vehicle track interaction measurement system of the diagnostic vehicle. Labeled information stemming from rail-head image-based detection are combined with regular acceleration measurements to allow the capture of variations in measurement conditions such as measurement speed and weather. Two classification approaches are compared in this work. The first classifier exploits Convolutional Neural Networks (CNNs) that are applied to FFT coefficients, which are computed from acceleration time-series data. The second classifier relies on a more classical machine learning approach, applied on features that are extracted from the acceleration time series, which are then classified using random forests (RFs). RFs are compared to CNNs, because RFs allow for root cause analysis, while CNNs perform well across many time-series classification tasks. Finally, the uncertainty of the acceleration metrics and of the ground-truth labels is analyzed, motivating the application of acceleration-based detection for improvement of rail condition monitoring.

11.2 Methodology

Machine learning algorithms have had numerous successful applications for classifying signals from accelerations. The traditional approach requires the extraction of representative features from the acceleration time series. These features are then used for classification. Kubera et al. [5] compare Support Vector Machines (SVMs) and Random Forests (RFs) to detect the change of vehicle speed using basic features computed from the time–frequency representation of audio data. Operational parameters, such as the vehicle speed, directly relate to the response of the axle and thus these parameters are important to be taken into account. The longitudinal level D0 (wavelength 1–3 m), D1 (wavelength 3–25 m), and D2 (wavelength 25–70 m) are geometric parameters defined by the railway norm EN13848-1 [6]. The longitudinal levels extracted from Axle Box Accelerations using integration and filtering techniques have been shown to be robust, speed independent, and repeatable indicators [2, 7]. The high frequency effects such as the vibrational modes of rail, wheel, and axle affect the wavelengths of under one meter. The representation of these high frequency effects in the time–frequency domain is obtained by decomposing the signals using DWT or STFT. Both the STFT and the DWT satisfy the property of invertibility, but the DWT unlike the STFT offers a high resolution in time and in frequency [8]. The condensation of a signal into a sparse representation of essential attributes is generally achieved by computing statistical features [9]. These features include minimum, maximum, and mean values, as well as standard deviation and higher statistical moments (skewness, kurtosis).

Decision Trees (DTs) and Random Forests (RFs) are classic machine learning algorithms, which yield interpretable classification results, but similarly to SVMs work best on a reduced set of parameters. RFs are ensemble models that aggregate several decision trees to achieve a more robust prediction than an individual DT. A semi-supervised interpretable machine learning framework for Sensor Fault detection, based on SVMs trained on numerous features extracted from acceleration time series, combined with the SHAP algorithm [10] was proposed by Martakis et al. [11]. The methodology proposed here employs RFs to classify the class label based on the previously described features (see also Fig. 11.2).

Deep Neural Networks have increasingly become popular in solving complex time-series classification tasks with the increase of time-series data availability [12]. Deep learning can achieve similar or better results while avoiding to perform the feature engineering required for machine learning methods such as DTs and RFs. This is not to say that feature engineering has no place within deep learning, as the injection of some form of prior knowledge or physics in a learning system is often beneficial in better capturing the governing dynamics [13]. However, the ability of a neural network to ingest data and extract useful representations on the basis of examples is what makes deep learning so powerful. CNNs have often been used in tasks relating the classification of time series. In [14] a CNN has been applied to raw axle box acceleration data for the detection insulated joints. The methodologies for classifying axle box acceleration signals using CNNs and RFs are illustrated in Fig. 11.2.

11.3 Case Study

11.3.1 Data Description

The acceleration data are labelled under three main categories: *welds*, *insulated joints (Insul)*, and *surface defects (SurfDef)*. The model is trained to differentiate between these three categories and a fourth one, named *no event (NoEvent)*. This category corresponds to baseline samples where no label was identified for the signal. The labeled acceleration events are created by merging the peaks in the acceleration signal to the labels stemming from deep learning models applied on rail-head images. Defects recognized in the image are not always crossed by the wheel. Such occurrences are filtered out during the matching of acceleration peaks to component labels. The features used for training include accelerations, longitudinal level $D0$ and $D1$ obtained from accelerations, vehicle speed, DWT wavelet coefficients (bior 2.2), and their stochastic analysis (mean, percentiles, number of 0 crossings, etc.). The RFs are trained on this reduced features set. The CNN models are trained on the STFT representation of the time series.

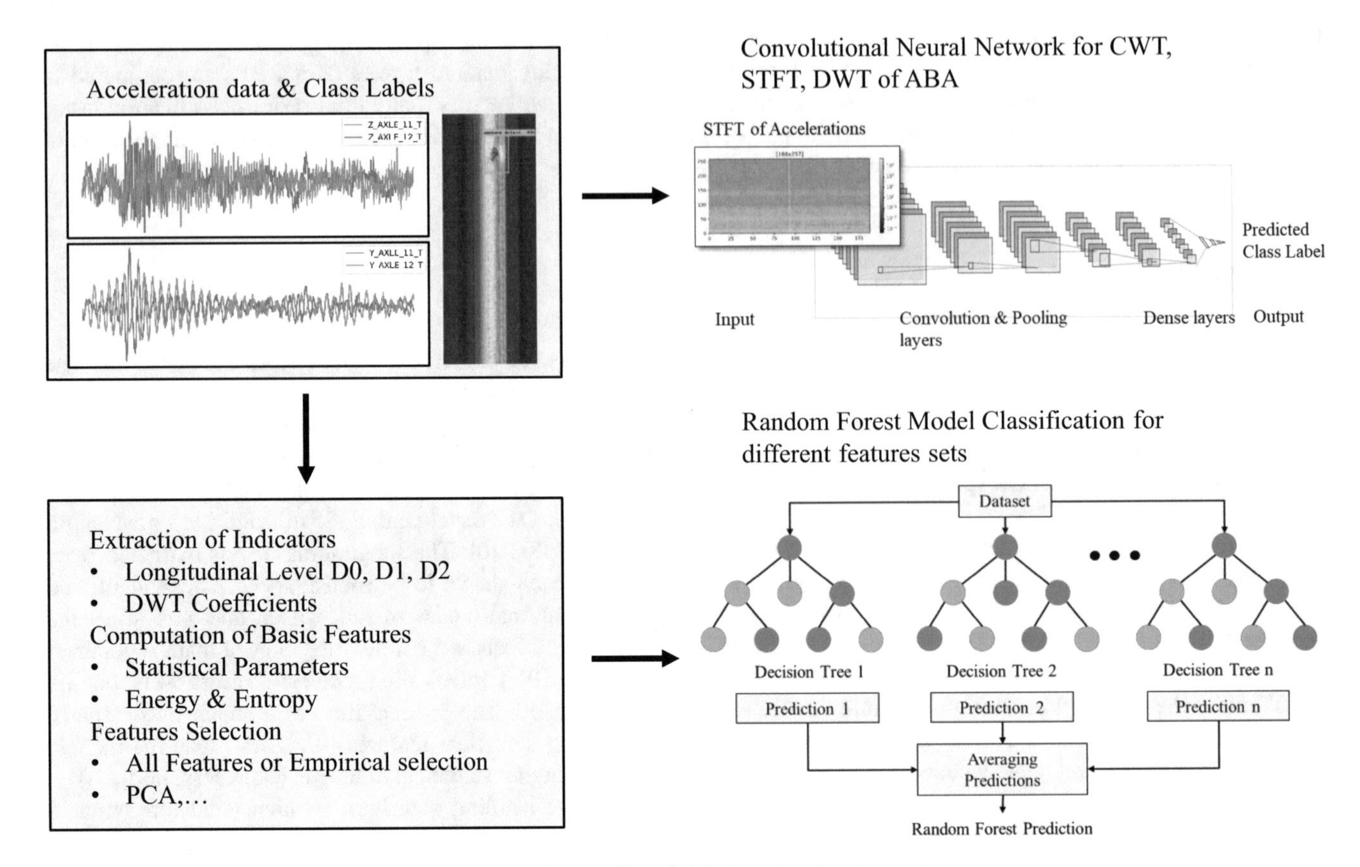

Fig. 11.2 Proposed methodologies using CNN and RF for classifying labeled acceleration time series

Table 11.1 Number of samples in the database

	No events	Welds	Insulated joints	Surface defects
Sample count	180,042	179,097	11,808	9657

The available data are highly imbalanced as the two classes *welds* and *no event* have a number of observations that are a magnitude higher than the ones of *insul* and *surface defects*. Indeed, when a track is split into equal segments, the number of segments presenting a defect is much smaller than the number of healthy segments. Insulated joints are less common than welds, as they are only chosen over welds when necessary. The samples count is summarized in Table 11.1. The data was subdivided into a training and a testing set. A balanced dataset is required in order to train the model for each category and not just for the most represented classes. Balanced datasets are obtained either by decreasing the class weights of the over-represented category or by oversampling the minority class via use of a Synthetic Minority Oversampling Technique (SMOTE) [15].

11.3.2 Classification Results

The RF and the CNN are trained using the data pre-processing steps illustrated in Fig. 11.2. The model performance is compared for different data preprocessing steps in Table 11.2. The confusion matrices are illustrated for the classifiers with the best scores for the RF classifier in Fig. 11.3 and for the CNN model in Fig. 11.4. RF- and CNN-based models result in similar metrics.

For both RF and CNN models, a test score of around 62–71% is achieved for the classification of component and damage labels using only preprocessed accelerations and vehicle speed as main inputs. The best RF model (71% average accuracy) was achieved by taking only the data from the accelerometers on the first axle raw data on which the following features were computed: Stochastic analysis of the accelerations, bandpass filtered signals D0 and D1, DWT of ABA (bior 2.2), and speed. Using Principal Component Analysis (PCA) to combine correlated features did not increase the model accuracy.

Table 11.2 Average accuracy scores of the most important models based on CNN and RF

Classifier	Preprocessing	Average accuracy
CNN	Speed normalized FFT	0.65
CNN	Speed normalized CWT	0.62
RF	Statistics on D0, D1 and DWT Bior 2.2 of ABA, speed	0.71
RF	PCA of statistics on D0, D1 and DWT Bior 2.2 of ABA, speed	0.64

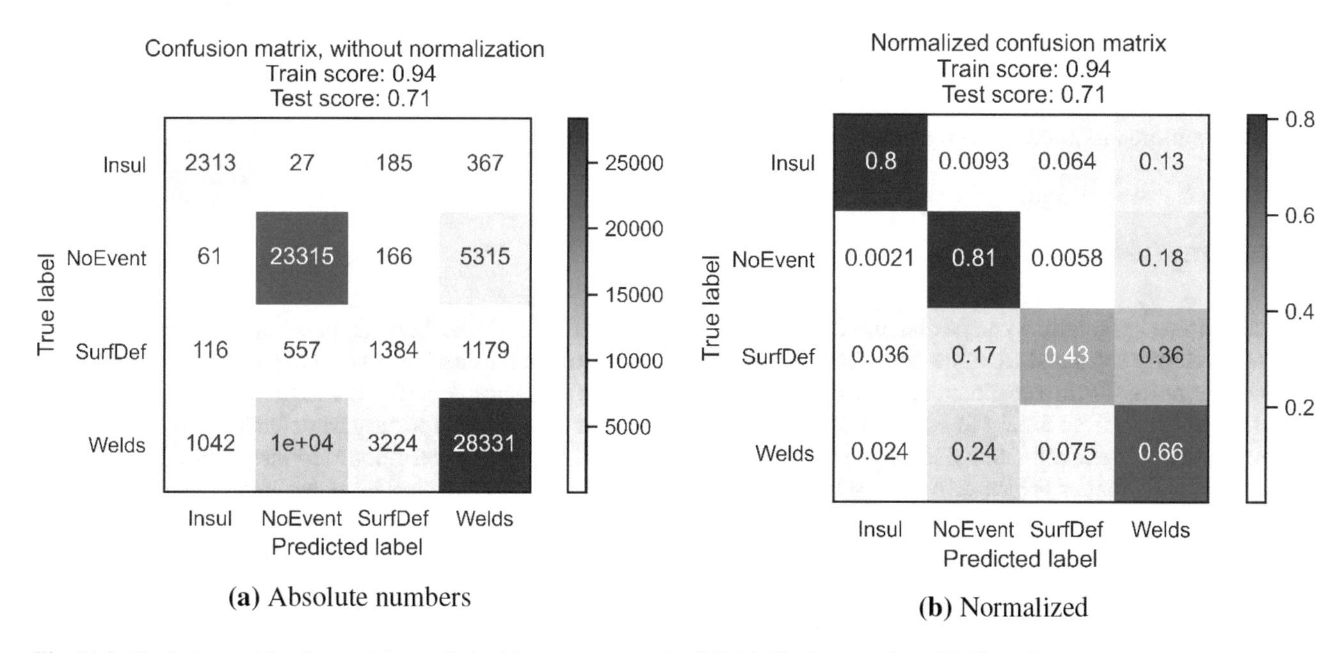

Fig. 11.3 Confusion matrix after applying optimized hyperparameters for RF. (**a**) Absolute numbers. (**b**) Normalized

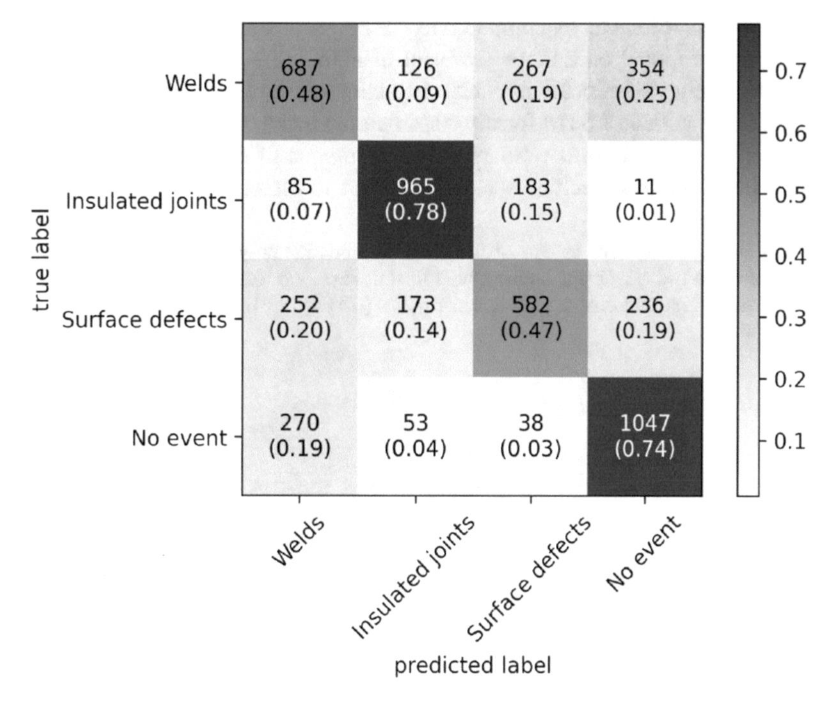

Fig. 11.4 Confusion matrix of CNN model on speed normalized STFT

STFT delivered the best classification results out of the different preprocessing approaches tested (STFT and CWT). The CWT and its subcategories scored similar results, although always slightly worse than the Fourier approach. Normalizing the data with the speed of the carriage, balancing the input data and increasing the number of train events were the steps that brought the major improvements to the results. Using only one axle instead of two, swapping the ABA data in order to have all defects on the same side, and changing the number of layers in the TensorFlow model only had a minor impact on the results.

For all models, it can be observed that there is always a trade-off between accuracy and precision: It is possible to get over 80% of the welds recognized correctly; however, in such a model many surface defects are also classified as welds. On the other hand, when the model has a good detection rate of the surface defects, the detection rate of the welds drops to around 60%. Insulated joints do not show such a clear trade-off as the acceleration response over this component can be separated quite easily from the other classes. In the conclusion, the sources of these trade-offs are explained and process and model improvements are proposed which may yield improved classification results.

11.4 Conclusions

The rail surface irregularities can be classified with a mean accuracy between 62 and 72%. Certain classes, such as insulated joints, are clearly separated from the other fault classes. The classification of surface defects, welds, and no events is more challenging, as they lie in a continuum of axle box acceleration response dynamics. A perfectly executed weld is, for instance, hard to distinguish from a flat rail surface (no event). Certain welds may assimilate to surface defects due to a surface or internal material irregularity. Smaller surface defects do not cause a significant acceleration response, while large surface defects cause a stronger response. A finer subdivision of categories, such as welds and insulated joints, into *healthy* and *damaged* component categories, is being explored as a refinement of the scheme. The labels used in the classification are first automatically assigned from images. The assigned labels may in some cases be erroneous (false positive or misclassified samples). The categorization of a defect is not always 100% clear and even when manually evaluating the image, there is some room for interpretation. The use of only human validated events may, at the cost of a lower sample size, help to filter out such unclear or erroneous instances. While the proposed models are successful at classifying accelerations, further improvements can be obtained by some extensions of the current framework. The combination of the RF model with the SHapley Additive exPlanations (SHAP) algorithm could be used to achieve a more interpretable classification framework [11]. The slightly lower performance of the CNN compared to RF may be explained by the smaller dataset used for training the model. Aside from training the model on a more powerful machine, directly feeding the untransformed measurement data into the CNN may result in better classification results. In such an instance, the CNNs would essentially act as feature detectors, which may be combined with an LSTM for the classification task [16]. The application of the classifiers to tracks with no prior knowledge is an essential step to validate the performance of data-driven machine learning models. These models may in future enable rail condition monitoring using low-cost accelerometers on OBM vehicles.

Acknowledgments This work is supported by SBB as part of the Mobility Initiative program under project OMISM—On-board Monitoring for Integrated Systems Understanding and Management Improvement in Railways. We would like to thank in particular our partners from the Metrology (MUD) and Strategic Asset Management of the Track departments (SAFB) of SBB.

References

1. UIC: UIC Code 712. Rail Defects, 4th edn., pp. 106–107. International Union of Railways, Paris (2002)
2. Dertimanis, V.K., Zimmermann, M., Corman, F., Chatzi, E.N.: On-board monitoring of rail roughness via axle box accelerations of revenue trains with uncertain dynamics (2019)
3. Ágh, C.: Comparative analysis of axlebox accelerations in correlation with track geometry irregularities. Acta Tech. Jaurinensis **12**, 161–177 (2019)
4. Zili, L., Molodova, M., et al.: Improvements in axle box acceleration measurements for the detection of light squats in railway infrastructure. Trans. Ind. Electron. **62**(7), 4385–4396 (2011)
5. Kubera, E., Wieczorkowska, A., Kuranc, A., Słowik, T.: Discovering speed changes of vehicles from audio data. Sensors (Switzerland) **19**, 3067 (2019)
6. CEN: En 13848-1, railway applications. Track. Track geometry quality. Characterization of track geometry. In: BSI (2019)
7. Hoelzl, C., Dertimanis, V., Landgraf, M., Ancu, L., Zurkirchen, M., Chatzi, E.: On-board monitoring for smart assessment of railway infrastructure: a systematic review. In: The Rise of Smart Cities: Advanced Structural Sensing and Monitoring Systems (2022)
8. Goswami, J.C., Chan, A.K.: Fundamentals of Wavelets: Theory, Algorithms, and Applications, 2nd edn. Wiley, New York (2010)

9. Rees, D.G.: Summarizing data by numerical measures (2020)
10. Thomson, W., Roth, A.E.: The Shapley value: essays in honor of Lloyd S. Shapley. Economica **58** (1991)
11. Martakis, P., Movsessian, A., Reuland, Y., Pai, S.G.S., Quqa, S., Garcia Cava, D., Tcherniak, D., Chatzi, E.: A semi-supervised interpretable machine learning framework for sensor fault detection. Smart Struct. Syst. Int. J. **29**, 251–266 (2022)
12. Ismail Fawaz, H., Forestier, G., Weber, J., Idoumghar, L., Muller, P.A.: Deep learning for time series classification: a review. Data Min. Knowl. Disc. **33**, 917–963 (2019)
13. Lai, Z., Mylonas, C., Nagarajaiah, S., Chatzi, E.: Structural identification with physics-informed neural ordinary differential equations. J. Sound Vib. **508**, 116196 (2021)
14. Yang, C., Sun, Y., Ladubec, C., Liu, Y.: Article developing machine learning-based models for railway inspection. Appl. Sci. (Switzerland) **11**, 13 (2021)
15. Chawla, N.V., Bowyer, K.W., Hall, L.O., Philip Kegelmeyer, W.: SMOTE: synthetic minority over-sampling technique. J. Artif. Intell. Res. **16**, 321–357 (2002)
16. Mylonas, C., Chatzi, E.: Remaining useful life estimation under uncertainty with causal GraphNets. arXiv:2011.11740 (2020)

Chapter 12
Development of a Surrogate Model for Structural Health Monitoring of a UAV Wing Spar

Adrielly H. Razzini, Iddo Kressel, Yoav Ofir, Moshe Tur, Tal Yehoshua, and Michael D. Todd

Abstract A critical part to implementing a structural health monitoring system is being able to understand the structural response under different operational and environmental conditions. In this work, a detailed finite element model of an unmanned aerial vehicle's wings' spar was developed to serve as a synthetic data generator. A probabilistic understanding of the aerodynamic loads and debonding damages at different locations and with different sizes were implemented to simulate observations of the spar's performance in service. The target measurements are uniaxial strain, measured in several paths throughout the spar. Given measured strain, the damage assessment problem is probabilistically formulated by defining local buckling from debonding as the observable damage, which is fundamentally characterized by load-dependent buckling eigenvalues. This FE physical model is highly computationally intensive, so a Gaussian process regressor and a multilayer artificial neural network (MANN) were designed to serve as a "run time" surrogate model to learn the relationships between inputs (loads and damage conditions) and outputs (strain measurements and buckling eigenvalues). The results illustrate that the surrogate models presented are a reliable replacement to the computationally expensive inverse finite element model in damage identification.

Keywords Structural health monitoring · Finite element · Surrogate model · Gaussian process regressors · Neural networks

12.1 Introduction

Modern unmanned aerial vehicles (UAVs) have high performance demands and are often subjected to extreme loading and environmental conditions. They are typically constructed from composite materials due to high strength-to-weight ratios [1], but such materials might suffer internal damage without any obvious changes on the surface, only becoming evident in the proximity of failure. According to aviation requirements, any damage that would prevent the aircraft from carrying its ultimate service load must be detected [2]. In this work, the structure being monitored is the wing spar of a UAV, and the observable damage was defined as debonding of the spar's top flange from the shear clips and web. As the wings deflect upward during flight and the top flange is compressed, this debonding can cause local buckling, which is critical to the structure. An FE model was developed to simulate hundreds of different combinations of damage size and location along the beam. Then, using this data for training, validation, and testing, machine learning was used to create an inexpensive surrogate model to learn the relationships between inputs (loads and damage conditions) and outputs (strain measurements and buckling eigenvalues).

The FE model was developed using ABAQUS CAE. The spar is 5 m long and is entirely made of carbon fiber-reinforced polymer composites, with a sandwich panel web. To reduce computational costs, 3D linear shell elements were used instead

A. H. Razzini · M. D. Todd (✉)
Department of Structural Engineering, University of California San Diego, La Jolla, CA, USA
e-mail: ahokamar@eng.ucsd.edu; mdtodd@eng.ucsd.edu

I. Kressel · Y. Ofir
Engineering Division, Israel Aerospace Industries (IAI), Ben Gurion International Airport, Tel-Aviv, Israel
e-mail: ikressel@iai.co.il; yoaofir@iai.co.il

M. Tur · T. Yehoshua
School of Electrical Engineering, Tel-Aviv University, Tel-Aviv, Israel
e-mail: tur@tauex.tau.ac.il; tal_yehoshua@mod.gov.il

R. Madarshahian, F. Hemez (eds.), *Data Science in Engineering, Volume 9*, Conference Proceedings of the Society for Experimental Mechanics Series, https://doi.org/10.1007/978-3-031-04122-8_12

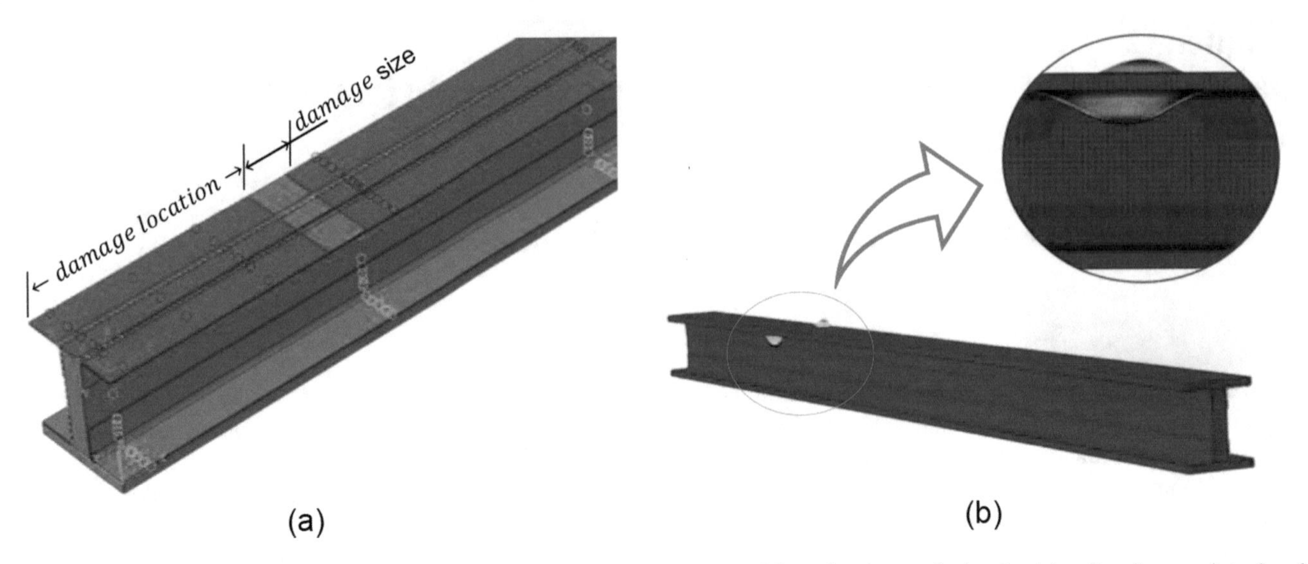

Fig. 12.1 Snapshots of the finite element model. (a) The top flange was suppressed from the view to display the debonding damage, introduced by removing the tie constraints. (b) Results of the buckling analysis, where local buckling occurred

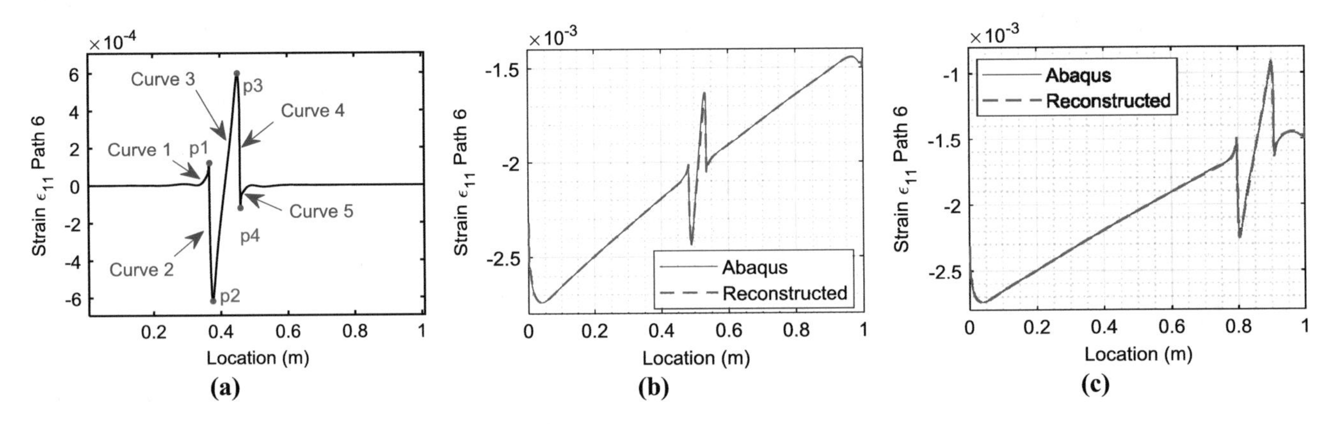

Fig. 12.2 (a) Strain measurement without baseline, divided into 5 curves that can be described by the coordinates of points p1–p4. (b, c) Strain measurement from Abaqus vs MANN reconstruction for different damage sizes and locations

of solid elements, and only a submodel of the first 1 m of beam was analyzed in this work. The damage is fully characterized by the damage parameters $\theta = \{\sigma, x\}$, where σ and x are independent random variables that refer to damage size and location, respectively. The debonding damage was introduced by removing a section of the bonding (tie constraints) between the top flange and the rest of the structure, as shown in Fig. 12.1a. Figure 12.1b shows a case where local buckling occurred as consequence of the debonding. The expected maximum service load was applied to the structure, and the analysis was divided into two independent steps – static, to measure uniaxial strain at several paths along the beam, and buckle, to provide buckling eigenvalues (λ). Prior probability density functions of the damage parameters θ were used to create a comprehensive database of 2500 different samples that were simulated on ABAQUS, using a Python script to handle the model changes and data storage. The simulation is very computationally expensive, as one case takes up to 15 min to run.

12.2 Strain Prediction Using an Artificial Neural Network

Bayesian inference can be used to produce a stochastic estimation of damage parameters, or even to choose an optimal sensor placement design considering Bayes risk. However, the likelihood $p(\varepsilon|\,\theta)$ needs thousands of samples to be properly evaluated, where ε are the strain measurements. Running the FE model to do so is prohibitively time consuming, so a multilayer neural network was created on MATLAB to act as a "run-time" surrogate model.

To do so, strain measurements were collected from a path below the top flange, and the baseline signal was removed from them. Then, they were divided in 5 curves, as shown in Fig. 12.2a, and general equations that fully describe the strain

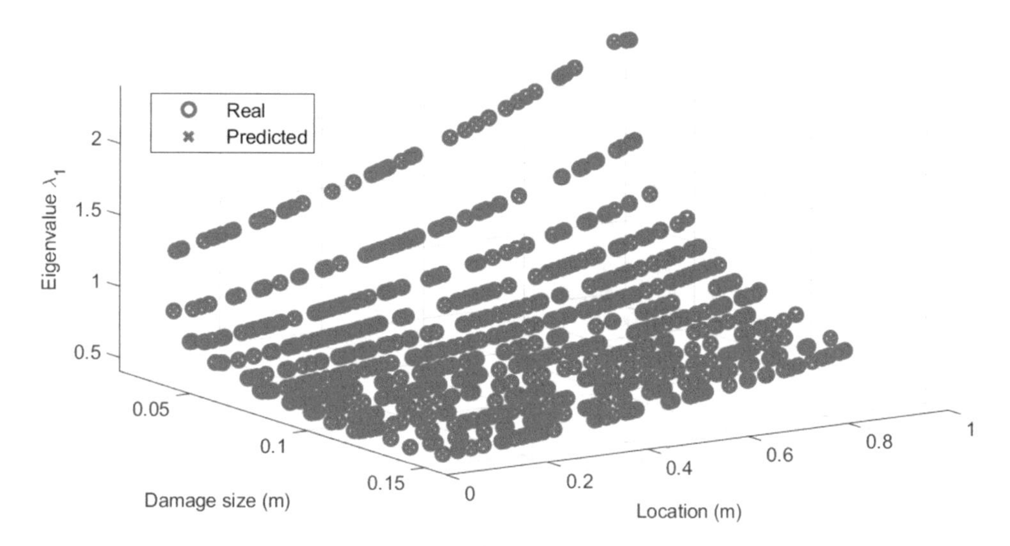

Fig. 12.3 Buckling eigenvalues obtained using Abaqus (blue) vs. predicted using the GPR model (orange)

signature based on the coordinates (x,y) of points p1 to p4 were found. These 8 coordinates are the target outputs from the regression MANN, and the two input features are damage size and location. Unlike normal regression where a single output value is predicted for each sample, multi-output regression requires specialized machine learning algorithms, such as deep neural networks, that support outputting multiple variables for each prediction. An advantage of these multi-output approaches is that they might produce simpler models with improved computational efficiency [3]. The number of hidden layers was defined first in a simple optimization. Then, a Bayesian optimization was performed to define the values of six hyperparameters of the NN. Samples of strain reconstruction using the 8 coordinates predicted by the trained NN are shown in Fig. 12.2b, c, and the results produced an average RMSE of 0.0067 for the testing data set. Evaluating a single sample takes around 10^{-3} s.

12.3 The Gaussian Process Regressor

Given that *failure load* $= \lambda \times$ *applied load*, the buckling eigenvalues can estimate the percentage of the applied loads the structure can withstand before the structure fails (i.e., local buckling occurs). This eigenvalue can be obtained using the FE model, but its high computational time makes it not suitable for real-time applications. Thus, a GPR was built as a surrogate model to learn the relationship between θ and λ. When compared to traditional NN models that only provide point estimates, the GPR models also provide standard deviations of the predictions that can be used to infer their expected accuracy [4]. A squared exponential kernel was used, and the 2500 samples were randomly divided in a 70/30 train/test split. The predictions using the GPR model yielded a RMSE = 0.00281, and the results for the testing data set are shown in Fig. 12.3.

12.4 Conclusion

This work presented a high-fidelity finite element model of an UAV wing's spar and introduced occasions in which using a surrogate model might be necessary. The results illustrate that the surrogate models presented had small RMSEs and are a reliable replacement to the computationally expensive inverse finite element model in damage identification. This approach directly supports modern model-assisted probability of detection approaches.

Acknowledgments This work is sponsored by the Israel Ministry of Defense under contract award #PO 4441026546.

References

1. Grodzki, W., Łukaszewicz, A.: Design and manufacture of umanned aerial vehicles (uav) wing structure using composite materials: Planung und bau einer flügelstruktur für unbemannte luftfahrzeuge (uav) unter verwendung von kompositwerkstoffen. Materialwissenschaft und Werkstofftechnik. **46**(3), 269–278 (2015)
2. North Atlantic Treaty Organization (NATO): AEP-4671 Unmaned Aircraft Systems Airworthiness Requirements (USAR), Edition B, Version 1, pp. 70–79. Nato Standardization Office (2019)
3. Kocev, D., Džeroski, S., White, M.D., Newell, G.R., Griffioen, P.: Using single-and multi-target regression trees and ensembles to model a compound index of vegetation condition. Ecol. Modell. **220**(8), 1159–1168 (2009)
4. Teimouri, H., Milani, A.S., Loeppky, J., Seethaler, R.: A Gaussian process–based approach to cope with uncertainty in structural health monitoring. Struct. Health Monit. **16**(2), 174–184 (2017)

Chapter 13
On a Description of Aeroplanes and Aeroplane Components Using Irreducible Element Models

Daniel S. Brennan, Robin S. Mills, Elizabeth J. Cross, Keith Worden, and Julian Gosliga

Abstract Previous work has shown that the damage classification performance for a particular Structural Health Monitoring problem on aeroplane components can be improved by using transfer learning. The transfer learning was aided by generating abstract representations of the components; these abstract representations are called *Irreducible Element* (IE) models. Such IE models have been applied previously for real-world bridge structures, encoding expert knowledge on the construction of bridges. By encoding such knowledge, it is hoped that similar IE models imply that the two structures (or subcomponents of said structures) are physically similar enough for positive data transfer, with the potential to improve the performance of damage classification on both structures. This chapter describes the process of generating an IE model for an entire aeroplane. By this process, rules of thumb for creating IE models, specific to aeroplanes, will be developed. Additionally, aeroplanes feature various materials, geometries, and functional components that are not seen in bridges. By attempting to describe an aeroplane, the list of valid geometric, material and contextual labels within the PBSHM is expanded.

Keywords Population-based structural health monitoring · Irreducible element model · Transfer learning · Aerospace · Knowledge transfer frameworks

13.1 Introduction

Population-based Structural Health Monitoring (PBSHM) is an approach to Structural Health Monitoring (SHM), which seeks to improve the ability to perform SHM by sharing data between structures [1–3]. To date, it has been shown that damage classification performance for a particular SHM problem on aeroplane wings can be improved by sharing data via transfer learning [4]. The two aircraft wings in question were taken from a Piper Tomahawk aircraft [5] and a Gnat trainer aircraft [6–8]. The first step was to create an IE model for both structures following the procedure outlined in [2]. This IE model largely ignored the complex geometry of the two wings with a focus instead on sensor locations, as the geometry was less relevant than the sensor locations for the SHM problem outlined in the chapter.

As PBSHM has been shown to be effective for at least one potential SHM problem in aerospace [4] and has been shown to be effective for grouping bridges by structural similarity [9], it is worth developing rules that allow IE models to be farmed for numerous aeroplanes and their components. In order to generate IE models for aerospace structures, problems such as the complex geometry of aerospace structures need to be overcome.

This chapter describes the procedure for generating a general IE model for a full size aircraft, namely a BAE Hawk T.Mk1 [10]. The term 'general IE model' is used here to refer to an IE model that is designed as a description of the structure in the absence of a specific SHM problem. The approach taken in this chapter is similar to the approach used for bridges in [9]. This chapter gives a brief overview of IE models in Sect. 13.2, Sect. 13.3 details how the BAE Hawk T.Mk1 is divided up into an IE Model with Sects. 13.3.1–13.3.3 covering Regular Elements, Relationships and Ground Elements, respectively.

D. S. Brennan · R. S. Mills · E. J. Cross · K. Worden · J. Gosliga (✉)
Dynamics Research Group, Department of Mechanical Engineering, University of Sheffield, Sheffield, UK
e-mail: dsbrennan1@sheffield.ac.uk; robin.mills@sheffield.ac.uk; e.j.cross@sheffield.ac.uk; k.worden@sheffield.ac.uk; j.gosliga@sheffield.ac.uk

R. Madarshahian, F. Hemez (eds.), *Data Science in Engineering, Volume 9*, Conference Proceedings of the Society for Experimental Mechanics Series, https://doi.org/10.1007/978-3-031-04122-8_13

13.2 Irreducible Element Models

The purpose of an IE model is to describe the physical nature of a structure for the purposes of comparing the similarity between two structures [2]. Describing the physical nature of a structure involves describing the overall topology of the structure (how different parts are connected), the materials found in each part and the geometry of each part of the structure. In this way, it is possible to tell whether two structures are similar, and where these similarities lie within the two structures. These aspects of the structure—geometry, materials and topology—are also believed to be important for ensuring knowledge is shared properly between structures via *transfer learning* [3].

A further label is added for each part that describes something of its functional nature, such as *aerofoil*, *column* or *deck*; this is called the *contextual label* and is the one part of the IE model that is not a purely physical description of the structure. Instead, it encodes engineering knowledge of how a particular component functions within a structure, which gives some implicit notion of the loads the component may experience, as well as other less tangible information. It is hoped that these contextual labels will make it easier to differentiate between different types of structures, for example a bridge will never contain an element labelled 'wing' (if it is designed properly), and therefore this could be used as a criteria to exclude bridge structures from a population of aeroplanes.

The IE model is produced by subdividing the structure into elements. This subdivision is largely dictated by the problem one is examining. For example, in the Gnat–Piper problem [4], the wings for each aircraft were subdivided into elements, such that the elements corresponded to particular damage labels. This subdivision was chosen as it allowed for damage localisation information to be transferred between the two structures. For more general descriptions, such as those found in [9], where the problem was to assess the overall similarity of bridges, the elements were chosen so that they each corresponded with a single structural component, i.e. each beam and column within the bridge was described with single elements.

13.3 Hawk Irreducible Element Model

The context for the IE model of the Hawk described in this chapter is one of assessing the Hawk's overall similarity to other aeroplanes. As such, the IE model discussed within this chapter will only focus on the overall airframe of the aeroplane, with each element representing a substantial geometrical object within the structure. While within the IE Schema [11], there is the flexibility to have every nut and bolt represented as an element, there is also introduced the idea of *granularity* of an IE model, which provides the flexibility to ignore the fine detail of a structure where it is not relevant to the problem at hand. If the context for generating the IE model were to change, and it was important to examine the structure in greater detail, it would be necessary for new IE models to be generated for the structure. For example, if one wished to locate damage in the hydraulics system, it would be necessary to add this into the IE model.

However, the context for generating the IE model of the Hawk in this chapter is to describe the overall geometry of the aircraft. Subsequently, the desired subdivision of elements will be one that allows for the best description of the overall geometry of the plane. Whereas for bridges, it was appropriate to describe each structural component with a single element because of their simple geometries; to accurately capture the complex geometries found in aircraft (especially jet aircraft), it was necessary to subdivide large components, such as the fuselage, into many smaller elements to enable both accurate representation of the geometrical shape, but also localisation of damage information within a structure.

Before the elements could be defined, it was useful to divide the aircraft up into major sections and components, namely: the body (see Fig. 13.1), consisting of the fuselage and vertical stabiliser; the aerofoils (see Fig. 13.2), consisting of both wings and both horizontal stabilisers; and finally, the landing gear (see Fig. 13.3), consisting of the left, centre and right landing gear. This is similar to the process used for bridges, where each bridge was divided into sections corresponding to the deck, the supports, the foundations, etc. Once these major components had been defined for the aircraft, the next step was to subdivide them into elements, such that the overall geometry could be accurately described.

These subdivisions of components are shown in Figs. 13.4, 13.5, and 13.6, with corresponding photos of the real aircraft provided for reference in Figs. 13.1, 13.2, and 13.3. Figures 13.1 and 13.4 depict how the fuselage and vertical stabiliser have been divided into their associated elements. The fuselage was subdivided along its length to match structural features visible on the surface of the aeroplane, while the vertical stabiliser has elements corresponding to major changes in curvature along the leading edge. Figures 13.2 and 13.5 show the subdivision of elements on both wings, as well as both horizontal stabilisers. The wings were subdivided along lines, which align with the control surfaces towards the rear of the wing, while the subdivisions of the horizontal stabiliser (like the vertical stabiliser) correspond to changes in the curvature along

Fig. 13.1 Side view of the Hawk T.Mk1 at the Laboratory for Verification and Validation

Fig. 13.2 Top down view of the Hawk T.Mk1 at the Laboratory for Verification and Validation

Fig. 13.3 Front view of the Hawk T.Mk1 at the Laboratory for Verification and Validation

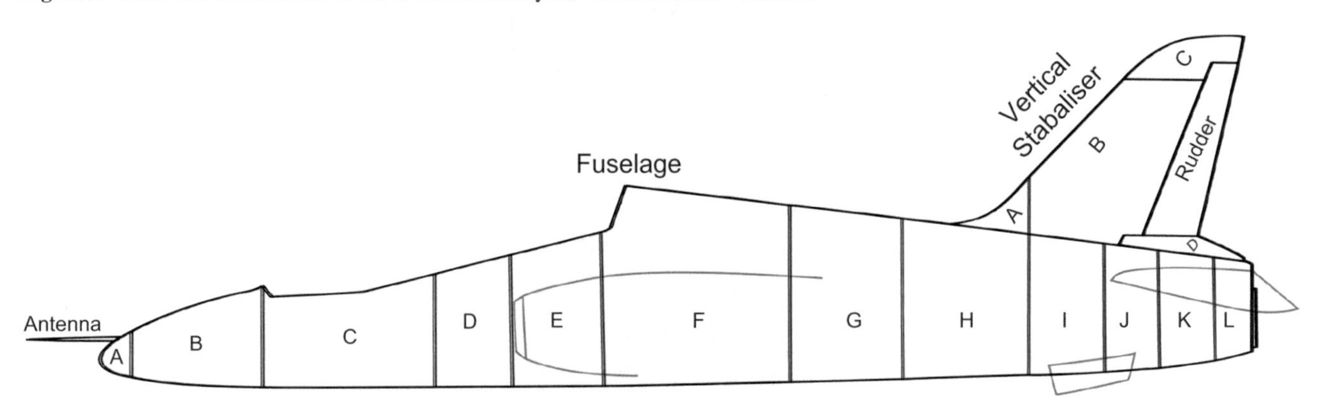

Fig. 13.4 Irreducible Element model breakdown of the centre section of the Hawk T.Mk1

the leading edge. Finally, Figs. 13.3 and 13.6 show the division of parts within the left, centre and right landing gear. The landing gear is subdivided into its main functional components.

It should be noted that the elements shown within this chapter are based on technical drawings by a well-respected illustrator [12]. Therefore, the IE model detailed within this chapter contains elements that are not present in our real reference Hawk T.Mk1. As such the IE model should be treated almost as an 'ideal' Hawk, and in comparison to the real plane, is almost more like the 'Form' of the Hawk [1]. Future work will involve generating an IE model of the actual plane in the lab, using measurements obtained from the real structure, and comparing this real example of a Hawk to the 'Form' generated here. This is a gross sort of damage detection itself.

In most cases, the subdivision of the major components was guided by the structural features/construction of plane. For example, where weld or rivet lines were present to follow, these were used to define the element boundaries. These 'naturally occurring' lines on the plane often corresponded to changes in the geometry of the major component. An example of this is the vertical stabiliser, where the rivet lines intersect the leading edge at points where there is a major change in the curvature of the leading edge. In other cases, the rivet lines on the aeroplane corresponded to some change in the internal structure that may be useful to capture in future iterations of the IE model. An example of this is between element G and element H in the

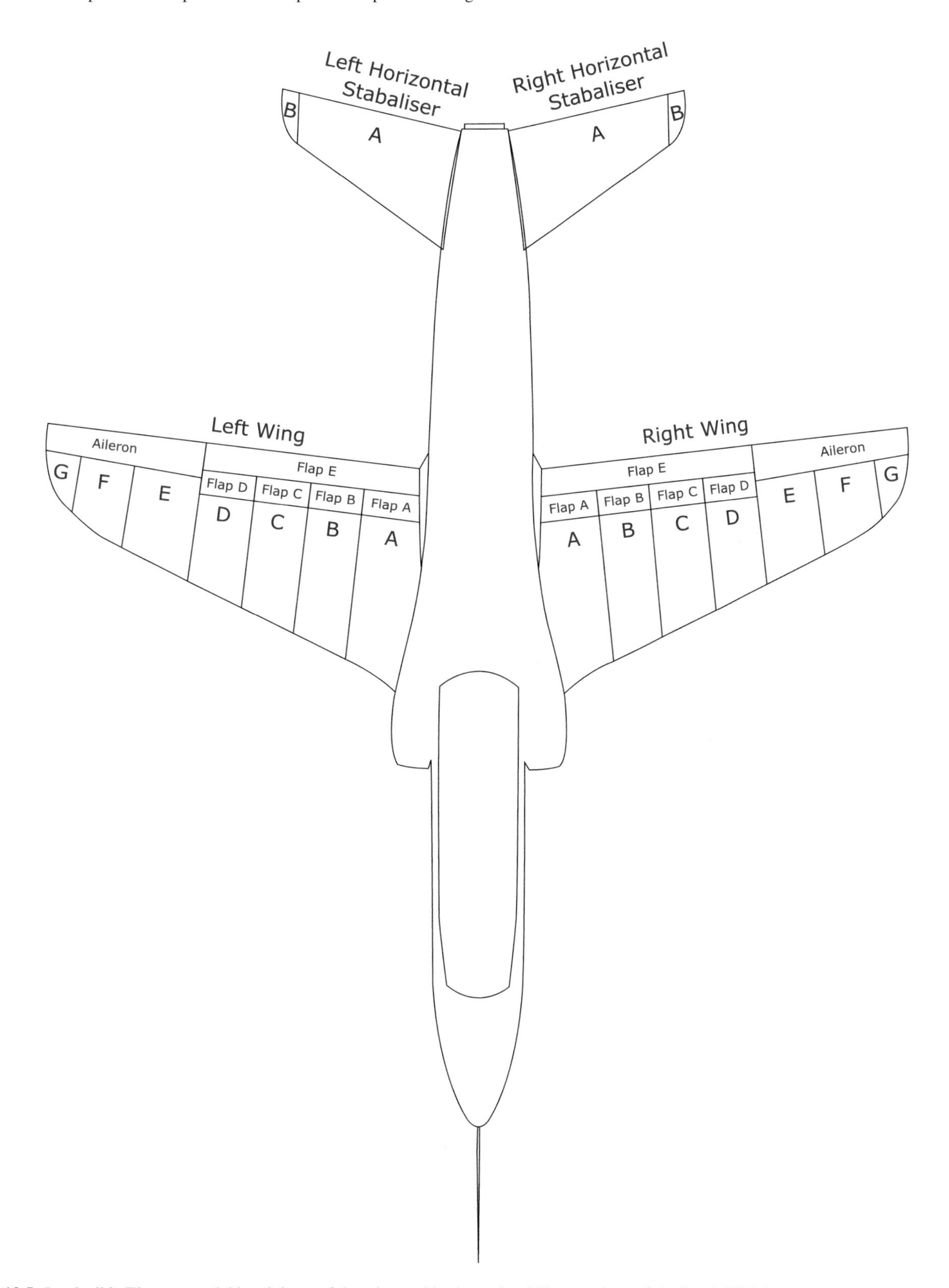

Fig. 13.5 Irreducible Element model breakdown of the wing and horizontal stabiliser sections of the Hawk T.Mk1

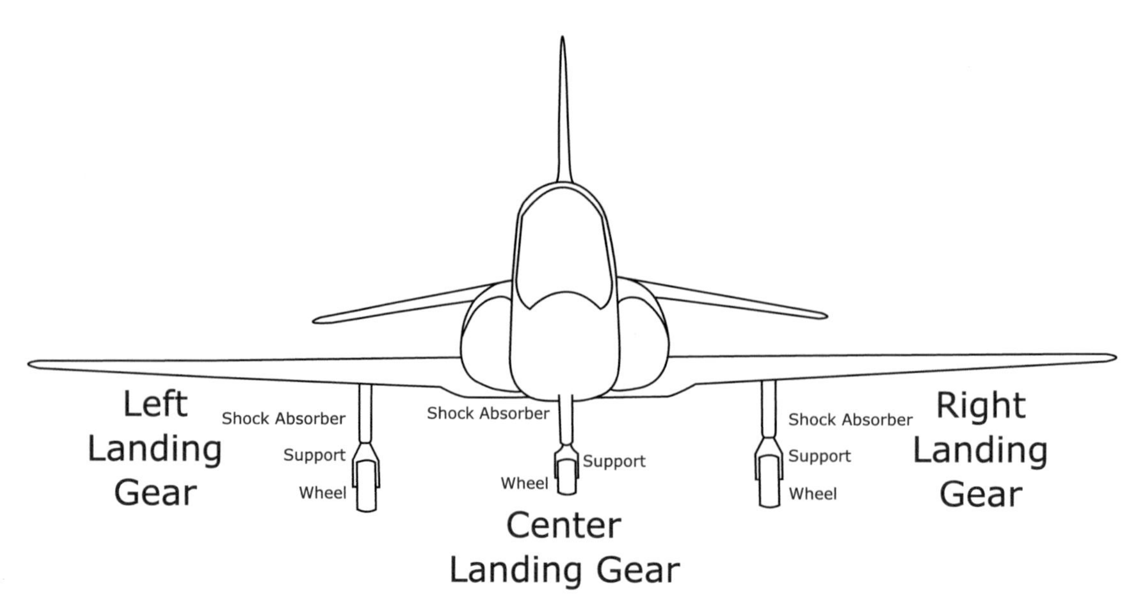

Fig. 13.6 Irreducible Element model breakdown of the landing gear of the Hawk T.Mk1

fuselage, where a rivet line on the underside corresponds to an internal bulkhead. Regardless, the actual construction of the plane provides a sensible guide for how the structure should be subdivided into elements.

In some cases, however, the subdivisions, while guided by the overall construction, do not necessarily correspond to construction indicators within the structure. Instead they correspond to direct 'characteristics' of the structure or the intersection where another section interacts with this 'characteristic'. An example of this is within the fuselage where it intersects with the horizontal stabilisers. One wishes to be able to localise damage on the intersection between the fuselage and the horizontal stabiliser; as such, an element K is defined to cover the geometrical area where this intersection occurs. Another example of this is in the wings, where the divisions between elements A, B, C and D are denoted. They no not correspond to any 'naturally occurring' lines on the wing, instead they correspond to the edges of flaps A, B, C and D to enable the IE model to accurately capture the 'nature' of the relationship between the wing division and its corresponding flap.

13.3.1 Materials, Geometry and Contextual Types

The material, geometric and contextual information for each element can be found in Tables 13.1, 13.2, and 13.3, with Table 13.1 corresponding to Figs. 13.1 and 13.4, Table 13.2 corresponding to Figs. 13.2 and 13.5 and Table 13.3 corresponding to Figs. 13.3 and 13.6.

The IE model of the Hawk presented in this chapter is by no means a perfect representation of the structure. Limitations in terms of the geometry that can be described in this current iteration of the schema [11], as well as the difficulty in obtaining detailed measurements of a structure this size, mean that at best this IE model is limited to describing the gross geometry of the aircraft.

Additionally, obtaining detailed material information for such an aircraft is difficult (especially considering the confidential nature of the aircraft's design). However, this is not necessarily important for the vast majority of SHM problems that will be concerned with the overall health of the aircraft. Particularly for identifying if two aircraft share similar failure modes, knowledge of whether or not a component is made of either an aluminium alloy or some form of composite should be sufficient. Therefore, for the purposes of this chapter, only a rough description of the material for each element will be included.

As mentioned previously, the contextual labels for each part describe something of its functionality, and it can be seen from Tables 13.1, 13.2, and 13.3 that these contextual labels largely correspond to the major components into which the aircraft was originally divided. This choice is because the major components all serve a functional purpose: the fuselage contains the engine and cockpit and provides a rigid structure; the wings provide lift; the stabilisers keep the plane aerodynamically stable; and the flaps provide control surfaces.

Table 13.1 Regular Elements for the fuselage and vertical stabiliser of the generalised Hawk T.Mk1

Element name	Contextual type	Material type	Geometry type
Fuselage			
Antenna	Other	Metal → Aluminium Alloy	Solid → Translate and Scale → Cylinder
Fuselage-a	Fuselage	Metal → Aluminium Alloy	Shell → Translate and Scale → Cylinder
Fuselage-b	Fuselage	Metal → Aluminium Alloy	Shell → Translate and Scale → Cylinder
Fuselage-c	Fuselage	Metal → Aluminium Alloy	Shell → Translate and Scale → Cylinder
Fuselage-d	Fuselage	Metal → Aluminium Alloy	Shell → Translate and Scale → Cylinder
Fuselage-e	Fuselage	Metal → Aluminium Alloy	Shell → Translate and Scale → Cylinder
Fuselage-f	Fuselage	Metal → Aluminium Alloy	Shell → Translate and Scale → Cylinder
Fuselage-g	Fuselage	Metal → Aluminium Alloy	Shell → Translate and Scale → Cylinder
Fuselage-h	Fuselage	Metal → Aluminium Alloy	Shell → Translate and Scale → Cylinder
Fuselage-i	Fuselage	Metal → Aluminium Alloy	Shell → Translate and Scale → Cylinder
Fuselage-j	Fuselage	Metal → Aluminium Alloy	Shell → Translate and Scale → Cylinder
Fuselage-k	Fuselage	Metal → Aluminium Alloy	Shell → Translate and Scale → Cylinder
Fuselage-l	Fuselage	Metal → Aluminium Alloy	Shell → Translate and Scale → Cylinder
Vertical stabiliser			
Vertical-stabiliser-a	Aerofoil	Metal → Aluminium Alloy	Shell → Translate and Scale → Cylinder
Vertical-stabiliser-b	Aerofoil	Metal → Aluminium Alloy	Shell → Translate and Scale → Cylinder
Vertical-stabiliser-c	Aerofoil	Metal → Aluminium Alloy	Shell → Translate and Scale → Cylinder
Vertical-stabiliser-d	Other	Metal → Aluminium Alloy	Shell → Translate and Scale → Cylinder
Rudder	Other	Metal → Aluminium Alloy	Shell → Translate and Scale → Cylinder

Table 13.2 Regular Elements for the right wing and horizontal stabiliser of the generalised Hawk T.Mk1. See Appendix 1 for the left counterpart

Element name	Contextual type	Material type	Geometry type
Right wing			
Right-wing-a	Wing	Metal → Aluminium Alloy	Shell → Translate and Scale → Cylinder
Right-wing-b	Wing	Metal → Aluminium Alloy	Shell → Translate and Scale → Cylinder
Right-wing-c	Wing	Metal → Aluminium Alloy	Shell → Translate and Scale → Cylinder
Right-wing-d	Wing	Metal → Aluminium Alloy	Shell → Translate and Scale → Cylinder
Right-wing-e	Wing	Metal → Aluminium Alloy	Shell → Translate and Scale → Cylinder
Right-wing-f	Wing	Metal → Aluminium Alloy	Shell → Translate and Scale → Cylinder
Right-wing-g	Wing	Metal → Aluminium Alloy	Shell → Translate and Scale → Cylinder
Right-flap-a	Aerofoil	Metal → Aluminium Alloy	Shell → Translate and Scale → Cylinder
Right-flap-b	Aerofoil	Metal → Aluminium Alloy	Shell → Translate and Scale → Cylinder
Right-flap-c	Aerofoil	Metal → Aluminium Alloy	Shell → Translate and Scale → Cylinder
Right-flap-d	Aerofoil	Metal → Aluminium Alloy	Shell → Translate and Scale → Cylinder
Right-flap-e	Aerofoil	Metal → Aluminium Alloy	Shell → Translate and Scale → Cylinder
Right-aileron	Aerofoil	Metal → Aluminium Alloy	Shell → Translate and Scale → Cylinder
Right horizontal stabiliser			
Right-horizontal-stabiliser-a	Aerofoil	Metal → Aluminium Alloy	Shell → Translate and Scale → Cylinder
Right-horizontal-stabiliser-b	Aerofoil	Metal → Aluminium Alloy	Shell → Translate and Scale → Cylinder

Table 13.3 Regular Elements for the right landing gear of the generalised Hawk T.Mk1. See Appendix 2 for the left and centre counterpart

Element name	Contextual type	Material type	Geometry type
Right landing gear			
Right-shock-absorber	Other	Metal → Aluminium Alloy	Solid → Translate → Cylinder
Right-support	Other	Metal → Aluminium Alloy	Plate → Other
Right-wheel	Wheel	Other	Other

There is a one component that is missing from the contextual labels, and that is the landing gear. That is because no contextual label for describing landing gear has yet been added to the list of valid contextual labels within the PBSHM database [13], where all the IE models will eventually be stored. The list of valid labels within the database is controlled by the database schema [11], and it is this schema that must be updated for the landing gear to be properly described with a contextual label.

Adding these labels when considering a new type of structure forms part of the research work on PBSHM. The current contextual types cover bridges well, as that has to date been the main focus of the work on generating IE models for structures. However, papers such as this one will eventually fill in the gaps for aeroplanes, and future papers will cover the contextual types for wind turbines and beyond.

As mentioned previously in this section, describing an aeroplane also requires geometric types that do not currently exist within the PBSHM database, so work must be done on storing descriptions for these new, complex geometries.

One type of geometry that is far more prevalent in aircraft than in bridges are shell geometries, which represent the fact that the airframe is largely hollow, often containing internal components. This type of geometry does not often exist in bridges in the same way. While there may still be hollow sections, such as box girders, these do not often contain other components.

13.3.2 Relationships Within the IE Model

Providing descriptions for the various elements is only the first part of the puzzle. Once the elements of the IE model have been defined, it is necessary to define how these elements relate to each another. For a complete and more detailed description of the various relationships that may exist within an IE model, the reader is directed towards [11].

These *relationships* can represent physical relationships, where it is theoretically possible to define the physical properties of the relationship. These are described as *joints* and are used to represent relationships such as welds or rivets, where one may eventually wish to model the physical behaviour of the joint. These joints fall into two major categories: static and dynamic. Static joints are joints where all degrees of freedom are restricted, and forces resist any relative motion. Dynamic joints on the other hand allow for relative motion in specific degrees of freedom, for example a hinge or some form of slider.

In contrast, certain relationships are purely abstract. The two abstract relationships featured in this chapter are the *perfect* relationship and the *boundary* relationship.

The perfect relationship represents where a larger component, for example a wing, is divided into elements to assist with solving a particular SHM problem. In this work, the wing was subdivided into elements more accurately capturing the geometry, but actually the elements together represent a single physical part of the structure. In the work on the transferring data from the Gnat wing to the Piper [4], the two wings were subdivided into elements that captured the damage locations of interest. The perfect relationship therefore provides the flexibility to further subdivide parts of the structure into elements to suit the requirements of the SHM problem, even where no physical division exists.

The boundary relationship represents the boundary of the IE model and must always connect to a ground element. Together the boundary relationship and ground element show where the physical description of an IE model ends, but where it is still connected to another physical structure that may or may not be modelled in future.

The relationships described in Table 13.4 relate to the elements described in Table 13.1. The relationships in Table 13.5 relate to the elements described in Table 13.2. The relationships described in Table 13.6 relate to both the elements described in Table 13.3 as well as the ground elements included in Table 13.7.

It should be noted that the description of the joints in these tables is vague, stopping at defining whether they are either static or dynamic, but providing no further physical description of the joint. This is because this is an active research area. Describing joints is as complex, if not more so, than describing the elements of a structure, and so much work remains to be done. However, for judging structural similarity, even knowing which joints are dynamic and which are static goes a long way to differentiating two structures.

13.3.3 Ground Elements

As mentioned in the previous section, the boundary relationships show which parts of the structure are related to the ground elements, listed in Table 13.7. In this particular case the ground is literally just that, the ground. However, in other work [4], the ground may be used to represent other structures that the IE model is attached to, but for the purposes of a particular SHM

Table 13.4 Relationships for the fuselage and vertical stabiliser of the generalised Hawk T.Mk1

Relationship name	Element set	Type
Fuselage		
Antenna-fuselage-a	{antenna, fuselage-a}	Joint → Static
Fuselage-a-b	{fuselage-a, fuselage-b}	Perfect
Fuselage-b-c	{fuselage-b, fuselage-c}	Perfect
Fuselage-c-d	{fuselage-c, fuselage-d}	Perfect
Fuselage-d-e	{fuselage-d, fuselage-e}	Perfect
Fuselage-e-f	{fuselage-e, fuselage-f}	Perfect
Fuselage-f-g	{fuselage-f, fuselage-g}	Perfect
Fuselage-g-h	{fuselage-g, fuselage-h}	Perfect
Fuselage-h-i	{fuselage-h, fuselage-i}	Perfect
Fuselage-i-j	{fuselage-i, fuselage-j}	Perfect
Fuselage-j-k	{fuselage-j, fuselage-j}	Perfect
Fuselage-k-l	{fuselage-k, fuselage-l}	Perfect
Fuselage to vertical stabiliser		
Fuselage-h-vertical-stabiliser-a	{fuselage-h, vertical-stabiliser-a}	Joint → Static
Fuselage-i-vertical-stabiliser-b	{fuselage-i, vertical-stabiliser-b}	Joint → Static
Fuselage-j-vertical-stabiliser-b	{fuselage-j, vertical-stabiliser-b}	Joint → Static
Fuselage-j-vertical-stabiliser-d	{fuselage-j, vertical-stabiliser-d}	Joint → Static
Fuselage-k-vertical-stabiliser-d	{fuselage-k, vertical-stabiliser-d}	Joint → Static
Fuselage-l-vertical-stabiliser-d	{fuselage-k, vertical-stabiliser-d}	Joint → Static
Vertical stabiliser		
Vertical-stabiliser-a-b	{vertical-stabiliser-a, vertical-stabiliser-b}	Perfect
Vertical-stabiliser-b-c	{vertical-stabiliser-b, vertical-stabiliser-c}	Perfect
Vertical-stabiliser-b-d	{vertical-stabiliser-b, vertical-stabiliser-d}	Joint → Static
Vertical-stabiliser-b-rudder	{vertical-stabiliser-b, rudder}	Joint → Dynamic

Table 13.5 Relationships for the right wing and horizontal stabiliser of the generalised Hawk T.Mk1. See Appendix 1 for the Left counterpart

Relationship name	Element set	Type
Right wing		
Right-wing-a-b	{right-wing-a, right-wing-b}	Perfect
Right-wing-b-c	{right-wing-b, right-wing-c}	Perfect
Right-wing-c-d	{right-wing-c, right-wing-d}	Perfect
Right-wing-d-e	{right-wing-d, right-wing-e}	Perfect
Right-wing-e-f	{right-wing-e, right-wing-f}	Perfect
Right-wing-f-g	{right-wing-f, right-wing-g}	Perfect
Right-wing-flap-a	{right-wing-a, right-flap-a}	Joint → Dynamic
Right-wing-flap-b	{right-wing-b, right-flap-b}	Joint → Dynamic
Right-wing-flap-c	{right-wing-c, right-flap-c}	Joint → Dynamic
Right-wing-flap-d	{right-wing-d, right-flap-d}	Joint → Dynamic
Right-wing-e-aileron	{right-wing-e, right-aileron}	Joint → Dynamic
Right-wing-f-aileron	{right-wing-f, right-aileron}	Joint → Dynamic
Right-wing-g-aileron	{right-wing-g, right-aileron}	Joint → Dynamic
Right-wing-a-flap-e	{right-wing-a, right-flap-e}	Joint → Dynamic
Right-wing-b-flap-e	{right-wing-b, right-flap-e}	Joint → Dynamic
Right-wing-c-flap-e	{right-wing-c, right-flap-e}	Joint → Dynamic
Right-wing-d-flap-e	{right-wing-d, right-flap-e}	Joint → Dynamic
Right-wing-a-fuselage-f	{right-wing-a, fuselage-f}	Joint → Static
Right horizontal stabiliser		
Right-horizontal-stabiliser-a-b	{right-horizontal-stabiliser-a, right-horizontal-stabiliser-b}	Perfect
Right-horizontal-stabiliser-a-fuselage-k	{right-horizontal-stabiliser-a, fuselage-k}	Joint → Dynamic

Table 13.6 Relationships for the right landing gear of the generalised Hawk T.Mk1. See Appendix 2 for the left and centre counterpart

Relationship name	Element set	Type
Right landing gear		
Right-shock-absorber-wing-b	{right-shock-absorber, right-wing-b}	Joint → Dynamic
Right-shock-absorber-support	{right-shock-absorber, right-support}	Joint → Dynamic
Right-support-wheel	{right-support, right-wheel}	Joint → Dynamic
Right-wheel-ground	{right-wheel, right-ground}	Boundary

Table 13.7 Ground Elements for the landing gear of the generalised Hawk T.Mk1

Element name
Right-ground
Left-ground
Centre-ground

problem can be ignored. If the SHM problem were to change, however, it may become necessary to replace these boundary conditions with further IE models.

13.4 Conclusions

Developing an IE model for an aeroplane (the Hawk) has highlighted the differences in approach required for generating IE models for aeroplanes as opposed to IE models for bridges or other structures. The major difference is that the geometry of aeroplanes tends to be more complex than that of bridges, with the cross-sections of components often varying along their length. In order to describe the geometry of the chosen aeroplane accurately, it was necessary to subdivide the major components of the aircraft (wings, fuselage, etc.); whereas in a bridge, the geometry of structural components such as the beams and columns is captured well by just a single element.

Describing the geometry of planes completely will also require new geometric types to be included in the IE model schema, as current geometric types are built around either simple polygons or cross-sections appropriate to bridges, such as I-beams. Future work will focus on developing more geometric types that can fully capture the geometry of aerospace components.

As well as additional geometric types, generating the IE models for aeroplanes will require more contextual types to describe components such as the landing gear. Again, these are components that are not found in bridges and therefore do not exist within the current system. Other contextual types, such as wing and aerofoil, were already included as it was believed these would be useful; however, it is impossible to fully predict all possible contextual labels that may be required, and so the process for generating a full list is one of discovery.

The current IE model represents a dimensionless form of the Hawk aircraft. In the future, exact IE models representing the particular Hawk aircraft available at the Laboratory for Verification and Validation in Sheffield will be developed. The IE model for the particular Hawk will feature what is essentially damage in the form of missing panels and will allow us to compare an actual realisation of the form to the form itself.

Acknowledgments The authors of this chapter gratefully acknowledge the support of the UK Engineering and Physical Sciences Research Council (EPSRC) via grant references EP/J016942/1, EP/K003836/2 and EP/S001565/1. The authors of this chapter also wish to thank Michael Dutchman for his help in generating the IE model for the Hawk, as well as Andrew Bunce and David Hester for their contributions to the development of IE models for bridges.

Appendix 1: Left Aerofoil Irreducible Element Model

Regular Elements

Element name	Contextual type	Material type	Geometry type
Left wing			
Left-wing-a	Wing	Metal → Aluminium Alloy	Shell → Translate and Scale → Cylinder
Left-wing-b	Wing	Metal → Aluminium Alloy	Shell → Translate and Scale → Cylinder
Left-wing-c	Wing	Metal → Aluminium Alloy	Shell → Translate and Scale → Cylinder
Left-wing-d	Wing	Metal → Aluminium Alloy	Shell → Translate and Scale → Cylinder
Left-wing-e	Wing	Metal → Aluminium Alloy	Shell → Translate and Scale → Cylinder
Left-wing-f	Wing	Metal → Aluminium Alloy	Shell → Translate and Scale → Cylinder
Left-wing-g	Wing	Metal → Aluminium Alloy	Shell → Translate and Scale → Cylinder
Left-flap-a	Aerofoil	Metal → Aluminium Alloy	Shell → Translate and Scale → Cylinder
Left-flap-b	Aerofoil	Metal → Aluminium Alloy	Shell → Translate and Scale → Cylinder
Left-flap-c	Aerofoil	Metal → Aluminium Alloy	Shell → Translate and Scale → Cylinder
Left-flap-d	Aerofoil	Metal → Aluminium Alloy	Shell → Translate and Scale → Cylinder
Left-flap-e	Aerofoil	Metal → Aluminium Alloy	Shell → Translate and Scale → Cylinder
Left-aileron	Aerofoil	Metal → Aluminium Alloy	Shell → Translate and Scale → Cylinder
Left horizontal stabiliser			
Left-horizontal-stabiliser-a	Aerofoil	Metal → Aluminium Alloy	Shell → Translate and Scale → Cylinder
Left-horizontal-stabiliser-b	Aerofoil	Metal → Aluminium Alloy	Shell → Translate and Scale → Cylinder

Relationships

Relationship name	Element set	Type
Left wing		
Left-wing-a-b	{left-wing-a, left-wing-b}	Perfect
Left-wing-b-c	{left-wing-b, left-wing-c}	Perfect
Left-wing-c-d	{left-wing-c, left-wing-d}	Perfect
Left-wing-d-e	{left-wing-d, left-wing-e}	Perfect
Left-wing-e-f	{left-wing-e, left-wing-f}	Perfect
Left-wing-f-g	{left-wing-f, left-wing-g}	Perfect
Left-wing-flap-a	{left-wing-a, left-flap-a}	Joint → Dynamic
Left-wing-flap-b	{left-wing-b, left-flap-b}	Joint → Dynamic
Left-wing-flap-c	{left-wing-c, left-flap-c}	Joint → Dynamic
Left-wing-flap-d	{left-wing-d, left-flap-d}	Joint → Dynamic
Left-wing-e-aileron	{left-wing-e, left-aileron}	Joint → Dynamic
Left-wing-f-aileron	{left-wing-f, left-aileron}	Joint → Dynamic
Left-wing-g-aileron	{left-wing-g, left-aileron}	Joint → Dynamic
Left-wing-a-flap-e	{left-wing-a, left-flap-e}	Joint → Dynamic
Left-wing-b-flap-e	{left-wing-b, left-flap-e}	Joint → Dynamic
Left-wing-c-flap-e	{left-wing-c, left-flap-e}	Joint → Dynamic
Left-wing-d-flap-e	{left-wing-d, left-flap-e}	Joint → Dynamic
Left-wing-a-fuselage-f	{left-wing-a, fuselage-f}	Joint → Static
Left horizontal stabiliser		
Left-horizontal-stabiliser-a-b	{left-horizontal-stabiliser-a, left-horizontal-stabiliser-b}	Perfect
Left-horizontal-stabiliser-a-fuselage-k	{left-horizontal-stabiliser-a, fuselage-k}	Joint → Dynamic

Appendix 2: Left and Centre Landing Gear Irreducible Element Model

Regular Elements

Element name	Contextual type	Material type	Geometry type
Centre landing gear			
Centre-shock-absorber	Other	Metal → Aluminium Alloy	Solid → Translate → Cylinder
Centre-support	Other	Metal → Aluminium Alloy	Plate → Other
Centre-wheel	Wheel	Other	Other
Left landing gear			
Left-shock-absorber	Other	Metal → Aluminium Alloy	Solid → Translate → Cylinder
Left-support	Other	Metal → Aluminium Alloy	Plate → Other
Left-wheel	Wheel	Other	Other

Relationships

Relationship name	Element set	Type
Left landing gear		
Left-shock-absorber-wing-b	{left-shock-absorber, left-wing-b}	Joint → Dynamic
Left-shock-absorber-support	{left-shock-absorber, left-support}	Joint → Dynamic
Left-support-wheel	{left-support, left-wheel}	Joint → Dynamic
Left-wheel-ground	{left-wheel, left-ground}	Boundary
Centre landing gear		
Centre-shock-absorber-fuselage-b	{centre-shock-absorber, fuselage-b}	Joint → Dynamic
Centre-shock-absorber-support	{centre-shock-absorber, centre-support}	Joint → Dynamic
Centre-support-wheel	{centre-support, centre-wheel}	Joint → Dynamic
Centre-wheel-ground	{centre-wheel, centre-ground}	Boundary

References

1. Bull, L.A., Gardner, P.A., Gosliga, J., Rogers, T.J., Dervilis, N., Cross, E.J., Papatheou, E., Maguire, A.E., Campos, C., Worden, K.: Foundations of population-based SHM, Part I: homogeneous populations and forms. Mech. Syst. Signal Process. **148**, 107141 (2020)
2. Gosliga, J., Gardner, P.A., Bull, L.A., Dervilis, N., Worden, K.: Foundations of population-based SHM, Part II: heterogeneous populations – graphs, networks, and communities. Mech. Syst. Signal Process. **148**, 107144 (2020)
3. Gardner, P.A., Bull, L.A., Gosliga, J., Dervilis, N., Worden, K.: Foundations of population-based SHM, Part III: heterogeneous populations – transfer and mapping. Mech. Syst. Signal Process. **148**, 107142 (2020)
4. Gardner, P., Bull, L.A., Gosliga, J., Poole, J., Gowdridge, T., Dervilis, N., Worden, K.: A population-based methodology for transferring knowledge between heterogeneous structures in SHM: damage localisation in a population of aircraft wings. Mech. Syst. Signals Process. **172**, 108918 (2022)
5. Barthorpe, R.J., Manson, G., Worden, K.: On multi-site damage identification using single-site training data. J. Sound Vib. **409**, 43–64 (2017)
6. Worden, K., Manson, G., Allman, D.J.: Experimental validation of a structural health monitoring methodology: part I. Novelty detection on a laboratory structure. J. Sound Vib. **259**(2), 323–343 (2003)
7. Manson, G., Worden, K., Allman, D.J.: Experimental validation of a structural health monitoring methodology: part II. Novelty detection on a gnat aircraft. J. Sound Vib. **259**(2), 345–363 (2003)
8. Manson, G., Worden, K., Allman, D.J.: Experimental validation of a structural health monitoring methodology: part III. Damage location on an aircraft wing. J. Sound Vib. **259**(2), 365–385 (2003)
9. Gosliga, J., Hester, D., Worden, K., Bunce, A.: On population-based structural health monitoring for bridges. Mech. Syst. Signals Process. **173**, 108919 (2022)
10. Fraser-Mitchell, H.: The Hawk story. J. Aeronaut. Hist. **3**, 1–120 (2013)
11. Brennan, D.S., Gosliga, J., Cross, E.J., Worden, K.: On implementing an irreducible element model schema for population-based structural health monitoring. In: IWSHM 2021 (2021)
12. Rolfe, M.: BAE Hawk T.Mk1 and sub-types (2015). https://drawingdatabase.com/wp-content/uploads/2015/04/bae-hawk.gif
13. Brennan, D.S., Wickramarachchi, C.T., Cross, E.J., Worden, K.: Implementation of an organic database structure for population-based structural health monitoring. In: IMAC XXXIX (2020)

Chapter 14
Input Estimation of Four-DOF Nonlinear Building Using Probabilistic Recurrent Neural Network

Soheila Sadeghi Eshkevari, Iman Dabbaghchian, Soheil Sadeghi Eshkevari, and Shamim N. Pakzad

Abstract Input estimation is an essential task with various applications, especially in nonlinear dynamic systems. The model-based input estimation approaches are not feasible solutions for problems without well-known behaviors. Recently, data-driven methods have shown promise in capturing subtle nonlinearities in various domains. In this study, we utilize a machine learning approach for input estimation of a four-degree-of-freedom nonlinear building while capturing uncertainty quantification in predictions which is helpful to analyze the accuracy of the results. This numerical case study for the shear frame building with elastic perfectly plastic springs is considered to evaluate the applicability of the proposed input estimation method to nonlinear dynamic systems. The performance of the network is evaluated on fifteen testing ground motions, and the input estimation is accomplished with high accuracy.

Keywords Input estimation · Nonlinear building · Data-driven methods

14.1 Introduction

Estimation of the input force based on the response of engineering systems, such as mechanical, structural, and aerospace systems, is of vast attention. Input estimation is a key factor in the computation and experimental implementation of either system identification or simulation and analysis [1–3]. In the case of structures subjected to a variety of environmental and stochastic load sources, measuring the exact input loads is impractical. Additionally, as an inverse problem, issues such as nonunique solutions and high condition numbers can increase the restraints. The collective effect of the applied load is usually modeled as a white Gaussian random process, which is not necessarily accurate in the analysis. In this context, the limit for most of the existing methods is the reliance on prior information of the dynamic system. Sanchez and Benaroya [4] presents a discussion and review about the input estimation methods.

The two model-based methods for estimating the input of multi-degree-of-freedom (MDOF) systems which are variations of Kalman filtering [5–8] and Gaussian process latent force modes (GPLFM) [9, 10] have shown promising and high-accuracy outcomes, although the baseline model of the structure is a requisite. Hence, model-free algorithms, such as deep learning models, are implemented as a fitting alternative to model-based methods [11, 12]. Fully data-driven methods are suitable for unknown systems where understanding the properties of the system is computationally expensive or impractical. In recent years, data-driven methods for the estimation of the dynamic response of nonlinear systems have been developing. Zhang et al. [13] proposed a long short-term memory (LSTM) architecture for the full-state response modeling of nonlinear systems. Simpson et al. [14] studied nonlinear normal modes (NNM) prediction using LSTM for a full-response estimation on a 20-degree-of-freedom system. Najera and Lee [15] evaluated response estimation via temporal convolutional network compared to autoregression methods. Sadeghi Eshkevari et al. [16] proposed an RNN architecture for full-state response estimation of nonlinear systems in which the mathematical operations of the RNN cell resemble the associated operations from Newmark's method for nonlinear dynamic simulation. The promising performance of this physics-based neural network enables training with smaller variable space and datasets.

S. S. Eshkevari (✉) · I. Dabbaghchian · S. N. Pakzad
Civil and Environmental Engineering Department, Lehigh University, Bethlehem, PA, USA
e-mail: sos318@lehigh.edu; imd220@lehigh.edu; snp208@lehigh.edu

S. S. Eshkevari (✉)
MIT Senseable City Laboratory, Cambridge, MA, USA
e-mail: ssadeghi@mit.edu

R. Madarshahian, F. Hemez (eds.), *Data Science in Engineering, Volume 9*, Conference Proceedings of the Society for Experimental Mechanics Series, https://doi.org/10.1007/978-3-031-04122-8_14

Data-driven methods in the application of inverse dynamic problems are also investigated recently to estimate the load from the response [3, 17, 18]. Additionally, they studied model uncertainties and concluded that uncertainty quantification is critical to the construction of input force. The nonlinearity and complexity of the physical dynamic systems require a design approach that accomplishes the input estimation with no baseline model or extra conditions. Sadeghi Eshkevari et al. [19] proposed a probabilistic recurrent neural network (RNN)-based input estimation framework for nonlinear systems and evaluated this framework on a numerical quarter-car model, a real-world building, and a real-world vehicle suspension system. In the present paper, the RNN framework is implemented to learn the nonlinear input-to-output transformation of dynamic systems presented in [19] for a four-DOF nonlinear case study to estimate ground motion and evaluate the proposed framework.

14.2 Methodology and Case Study

Recurrent neural architectures have been widely proposed for signal regression tasks due to their capability of learning temporal dependencies and dynamic equations. In this paper, the main objective is to tackle the inverse problem: given y_k, the full state vector of the system at time step k, for $k \in 1:T$, and a prior estimate for u_T, the applied load at time step k, it is desired to estimate u_k for $k \in 1:T-1$. This problem is equivalent to the response deconvolution of a dynamic system (linear or nonlinear) without using the prior knowledge about the system. In this framework, the neural network is represented as an RNN block; at each time step, the RNN block processes the input and output values inside the binder to predict the one-step backward estimation of the input. This process is repeated until the maximum possible length of the input signal is estimated. In this framework, the input signals are associated to the ground motion of the building system, and story accelerations are systems' outputs.

The proposed methodology according to [19] consists of two steps: training and inference. In the training phase, multiple input and output signals from the structure are required so that the dynamical system can be learned by the RNN model. This is made possible by using finite element surrogate models for simulation. In the inference stage, however, the only input value that should be available is the input at the terminal state (the systems' input at the last discrete value of the signal). This input value in many applications can be simply set to zero considering an at-rest condition at the end of the sensing period (e.g., buildings after an earthquake will return to zero acceleration). Given this trivial input state, the system can unravel the previous inputs by processing the outputs that are fully available.

To train the proposed probabilistic regression model, conventional loss functions are not ideal since these functions incorporate deterministic values rather than distributions. Instead, the loss function has to directly incorporate the negative log likelihood of the observations given the model parameters [20]. In this context, the RNN block is parameterized by θ, and the goal is to maximize the probability of correctly estimating targets y_i given system inputs x_i under the trained parameters. With this definition, the loss function $L(\theta)$ is defined as the following:

$$L(\theta) = -\log\left(p\left(y_i | x_i, \theta\right)\right) + L_{\text{proj}}\left(\theta, n_{\text{proj}}\right) \tag{14.1}$$

where, $p(y_i | x_i, \theta)$ is the probability of drawing system input y_i given system output x_i and model parameters θ. The second term of the loss function is the projection loss with a projection length of n_{proj} (defined in [16]) and based on the parameterized model $f(\theta)$. As the likelihood term becomes smaller, we ensure that the network's output distributions are more likely to predict values that are close to the actual outputs. The second term of the loss function also attempts to enhance the regression accuracy for longer projections in a conventional mean squared error (MSE) minimization manner. In this term, a strictly increasing geometric factor is element-wise multiplied to the outputs in the trajectory to put more weight on the accuracy of more distant estimations.

To train the network using this loss function, Newton trust-region approach is adopted due to its superior performance compared to linear methods [21]. The training process consists of two phases: 50 epochs with projection length of five and 20 epochs with projection length of 50. To avoid overfitting, the performance of the models is evaluated using the cross-validation technique. The inclusion of the second term in the loss function is effective in preventing overfitting for its emphasis on long-run projected signals rather than immediate estimations. In addition, the compact design of the network helps to avoid overfitting by reducing the model variance. The optimal network designed for this case study includes five fully connected layers with Leaky ReLU activation function at all layers. After training phase, this network is developed to produce probabilistic excitation input in terms of mean value and standard deviation for uncertainty evaluation on the estimated values, given the structural response.

Table 14.1 Mechanical properties for the nonlinear building

Mechanical properties	Values	Units
M_1	0.259	kip.s^2/in
M_2/M_1	1	–
M_3/M_1	0.75	–
M_4/M_1	0.5	–
F_y	50	kips
k_1	168	kips/in
k_2/k_1	7/9	–
k_3/k_1	1/3	–
k_4/k_1	1/4	–

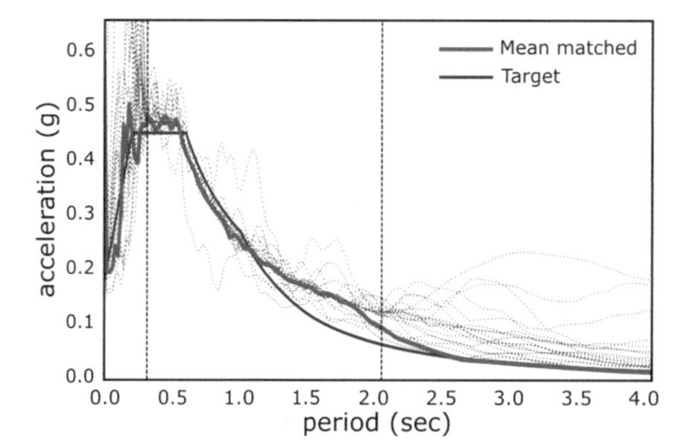

Fig. 14.1 Earthquake response spectra matched with respect to the target spectrum and the mean spectrum

In the following, we demonstrate efficacy of the proposed framework on a case study. In this numerical case study, a four DOF shear frame with elastic perfectly plastic springs is considered to evaluate the applicability of the proposed input estimation method to nonlinear dynamic systems. In this case study, the single input channel captures the ground motion and the response at each level of the shear frame constitutes the output channels. Equations 14. 2 and 14.3 and Table 14.1 display the equation of motion and mechanical properties for this system. It also presents the mass (M) and stiffness (K) from DOF 1 to DOF 4 (ascending with each floor of the shear frame) and the yielding force (F_y).

$$m\ddot{x} + c\dot{x} + f(x) = -m\Gamma\ddot{x}_g \tag{14. 2}$$

$$f_i(x) = \begin{cases} k_i x & x \leq \Delta_y \\ F_y & x > \Delta_y \end{cases} \tag{14.3}$$

Ground motion data identical to [16] is used for training and testing. In summary, a nonlinear Newmark's method in MATLAB is used to generate the response at each DOF for 30 ground motions. These ground motions are collected from the Center for Engineering Strong Motion Data (CESMD) within which 10 are generated in accordance with a band-limited white noise. All ground motions are then scaled following the wavelet method proposed by [22]. Figure 14.1 depicts the spectra for each ground motion, as well as the mean spectrum that closely follows the target spectrum. Lastly, the 30 ground motions are split evenly for training and testing. In this case study, the input of the RNN block includes a slice of the response at each DOF and one value of the ground motion from the current step. The slice length is a hyperparameter of the proposed method. Based on our preliminary trials, a slice length of 10 is found to outperform other values.

The performance of the network is evaluated on the 15 testing ground motions. Figures 14.2 and 14.3 show the time series and power spectral density (PSD) of the input estimation for a random test signal, respectively. In each time step, the resulted standard deviations from the neural network are useful to detect low confidence predictions in a local scale. Figure 14.2 shows high accuracy even at the largest amplitudes at $t = 17$ s and high confidence seen by the narrower confidence intervals.

Figure 14.4 compares the histograms of standard deviations in predicting inputs of a test signal with different noise levels. In this experiment, the response of the trained network to the same data with varying noise levels is experimented. The figure shows that, in general, the standard deviations move toward higher values as the noise level increases (e.g., signal-to-noise ratio (SNR) 5 is the most right-shifted case, and SNR 100 is the closest to the origin); however, the variation is quite small.

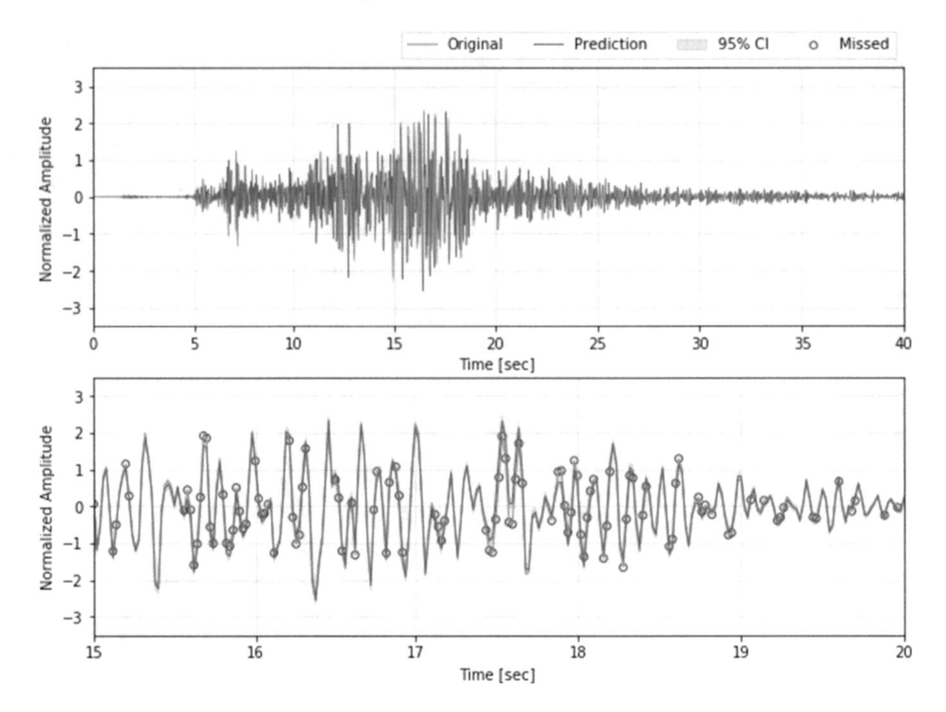

Fig. 14.2 Input signal prediction (~40 s)

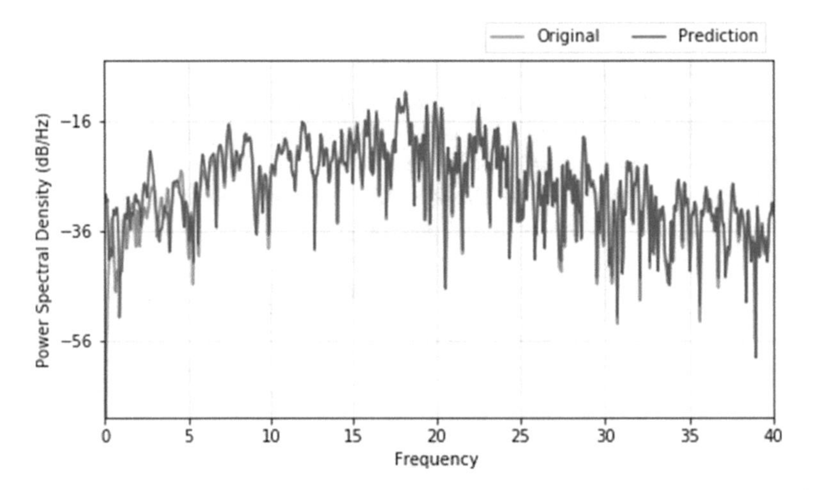

Fig. 14.3 PSD of input signal prediction

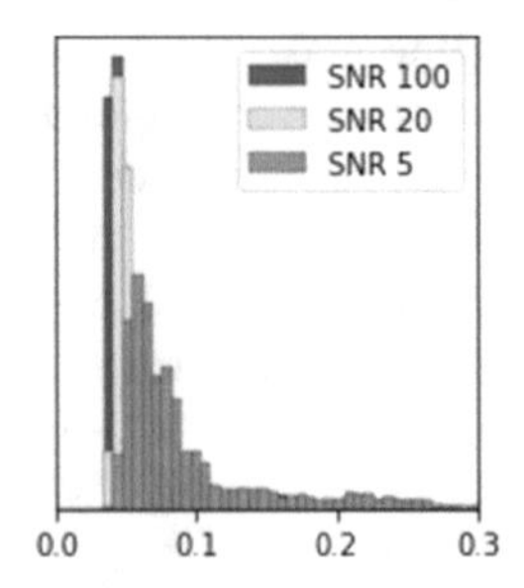

Fig. 14.4 Standard deviation histograms for signals with various SNRs

Therefore, as expected the network loses confidence in predicting inputs for noisier cases, but in a controlled level. Furthermore, from Fig. 14.2, the missed points, where the prediction is outside the 95% confidence interval, are all close to the reference; however, the confidence interval is narrow. The PSD in Fig. 14.3 reflects this high accuracy with the strong majority of the frequency spectrum matching the reference. However, it is important to note that the network underperforms for vibration frequencies below 5.0 Hz in this instance.

Table 14.2 Statistical properties derived from histograms of 50 × 12 random projections

Statistical property	Mean	Median
Correlation coefficient	0.9332	0.9380
MSE in time	0.0112	0.0112
MAE in frequency	1.5765	1.6469

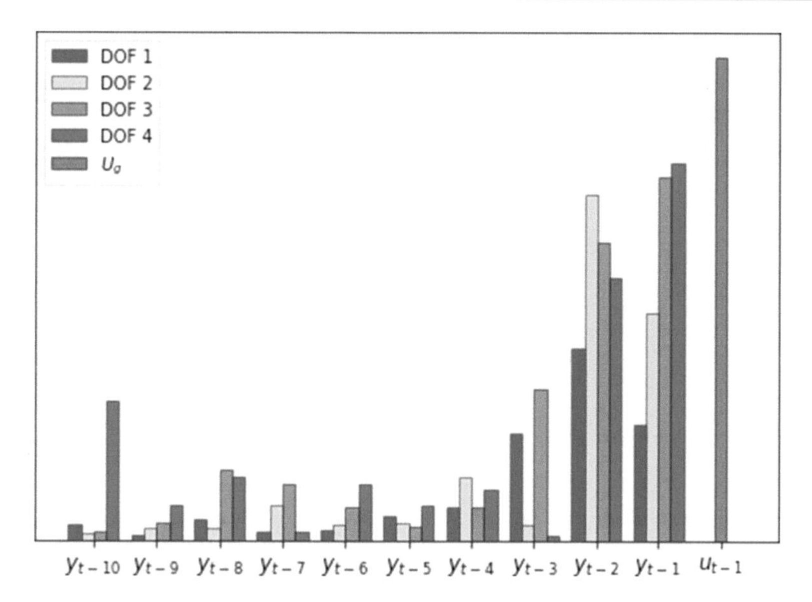

Fig. 14.5 Importance map of the network features

To quantify the input estimation performance of the trained network for all testing signals, the network is inferred repeatedly with the different testing earthquakes, random starting points, and projection lengths. For each trial, the original and the estimated signals are scaled by the maximum absolute value in the original signal for a fair comparison. The accuracy measures of all these test cases are presented in Table 14.2. The correlation coefficient is a good measure to quantify the general coherence of two signals. However, for more detailed comparison, mean squared error (MSE) of signals in time domain and mean absolute error (MAE) of signals in frequency domain are presented as histograms as well. The results in Table 14.2 demonstrate high accuracy with correlation coefficients near one and MSE near zero for the time series.

Learning-based solutions for engineering problems are often being criticized for their lack of interpretability. In recent years, deep learning community has been focused on developing fair methods to interpret and attribute the outputs of the network directly to its input features. Doing so, one can analyze whether the trained network make predictions by heavily relying on theoretically related features or not. In this study, we use integrated gradients which represent the integral of gradients with respect to the inputs throughout the network's depth from a given output back to the input layer [23]. The final product of this method is a relative importance map of the input features with respect to each output as presented in Fig. 14.5. It displays the contributions that each input has on the single output of this network. Inputs (structural response) in time proximity to the prediction's time step contribute substantially more than inputs at distant times, which is expected. In this case, due to the nonlinearity of the system, some inputs at distant times are also contributing noticeably, while this is not expected in linear systems.

To evaluate the performance of the model in uncertainty quantification, Fig. 14.6 is presented. This scatter plot shows residuals with respect to their predicted standard deviations. Residual captures the difference between the actual input value and the predicted value. The interesting finding is the monotonic relationship between the model error (residuals) and the model-predicted standard deviations. This implies that the model has learned to return higher uncertainty for low-accuracy estimations which has been the actual purpose of the neural architecture design.

14.3 Conclusion

This research evaluates a data-driven system input estimation approach with a nonlinear four-DOF case study. To investigate this method, this structure is considered subjected to 30 ground motions as the input excitation, split equally for training and testing of the probabilistic neural network. This RNN-based framework requires the structural response along the loading

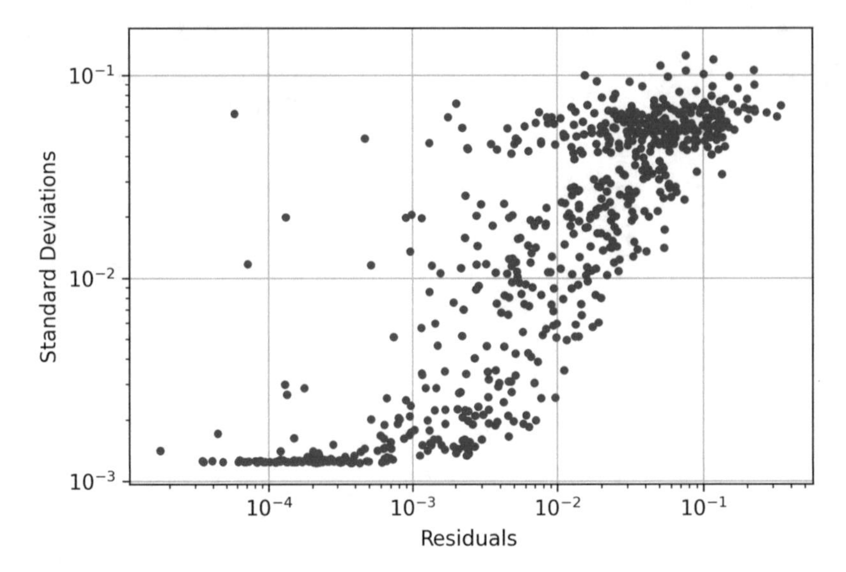

Fig. 14.6 Evaluation of uncertainty quantification

time history. Each prior time step excitation can be unraveled knowing a slice of previously experienced responses at each DOF along with the most recent input value. This framework is designed for a probabilistic prediction of the input load to the structure which is presented by mean and standard deviation values.

The comparison between the estimated and real input data in the frequency and time domains shows high accuracy and high confidence. The response with varying noise levels is also investigated, and, as the noise increases, the dispersion of the estimation would enlarge although the impact on the estimated signal is not significant. An importance map is also comparing the contribution of each DOF time step response on the input estimation objective. The nonlinearity of system can be observed by participation of prior time steps on the loading estimation. An evaluation of prediction error against predicted standard deviation also shows that probabilistic framework performs as expected.

References

1. White, J.R., Adams, D.E., Rumsey, M.A., van Dam, J., Hughes, S.: Impact loading and damage detection in a carbon composite TX-100 wind turbine rotor blade. In: 46th AIAA Aerospace sciences meeting and exhibit (2008). https://doi.org/10.2514/6.2008-1349
2. Park, J., Ha, S., Chang, F.-K.: Monitoring impact events using a system-identification method. AIAA J. **47**(9), 2011–2021 (2012). https://doi.org/10.2514/1.34895
3. Sarego, G., Zaccariotto, M., Galvanetto, U.: Artificial neural networks for impact force reconstruction on composite plates and relevant uncertainty propagation. IEEE Aerosp. Electron. Syst. Mag. **33**(8), 38–47 (2018). https://doi.org/10.1109/MAES.2018.170157
4. Sanchez, J., Benaroya, H.: Review of force reconstruction techniques. J. Sound Vib. **333**(14), 2999–3018 (2014). https://doi.org/10.1016/J.JSV.2014.02.025
5. Gillijns, S., de Moor, B.: Unbiased minimum-variance input and state estimation for linear discrete-time systems with direct feedthrough. Automatica. **43**(5), 934–937 (2007). https://doi.org/10.1016/J.AUTOMATICA.2006.11.016
6. Lourens, E., Reynders, E., de Roeck, G., Degrande, G., Lombaert, G.: An augmented Kalman filter for force identification in structural dynamics. Mech. Syst. Signal Process. **27**(1), 446–460 (2012). https://doi.org/10.1016/J.YMSSP.2011.09.025
7. Maes, K., Smyth, A.W., de Roeck, G., Lombaert, G.: Joint input-state estimation in structural dynamics. Mech. Syst. Signal Process. **70–71**, 445–466 (2016). https://doi.org/10.1016/J.YMSSP.2015.07.025
8. Eftekhar Azam, S., Chatzi, E., Papadimitriou, C.: A dual Kalman filter approach for state estimation via output-only acceleration measurements. Mech. Syst. Signal Process. **60–61**, 866–886 (2015). https://doi.org/10.1016/J.YMSSP.2015.02.001
9. Nayek, R., Chakraborty, S., Narasimhan, S.: A Gaussian process latent force model for joint input-state estimation in linear structural systems. Mech. Syst. Signal Process. **128**, 497–530 (2019). https://doi.org/10.1016/J.YMSSP.2019.03.048
10. Rogers, T.J., Worden, K., Cross, E.J.: On the application of Gaussian process latent force models for joint input-state-parameter estimation: with a view to Bayesian operational identification. Mech. Syst. Signal Process. **140**, 106580 (2020). https://doi.org/10.1016/J.YMSSP.2019.106580
11. Brunton, S.L., Proctor, J.L., Kutz, J.N.: Discovering governing equations from data by sparse identification of nonlinear dynamical systems. Proc. Natl. Acad. Sci. **113**(15), 3932–3937 (2016). https://doi.org/10.1073/PNAS.1517384113
12. Raissi, M., Perdikaris, P., Karniadakis, G.E.: Physics-informed neural networks: a deep learning framework for solving forward and inverse problems involving nonlinear partial differential equations. J. Comput. Phys. **378**, 686–707 (2019). https://doi.org/10.1016/J.JCP.2018.10.045

13. Zhang, R., Chen, Z., Chen, S., Zheng, J., Büyüköztürk, O., Sun, H.: Deep long short-term memory networks for nonlinear structural seismic response prediction. Comput Struct. **220**, 55–68 (2019). https://doi.org/10.1016/J.COMPSTRUC.2019.05.006
14. Simpson, T., Dervilis, N., Chatzi, E.: On the use of nonlinear normal modes for nonlinear reduced order modelling. In: Proceedings of the international conference on Structural Dynamic, EURODYN, vol 2, pp 3865–3877, 2020. Accessed 15 Oct 2021 [Online]. Available: https://arxiv.org/abs/2007.00466v1
15. Najera, D., Lee, C.: Structural response prediction of a flat beam with geometric nonlinearities with temporal convolutional networks. https://doi.org/10.31224/OSF.IO/9R3GX
16. Sadeghi Eshkevari, S., Takáč, M., Pakzad, S.N., Jahani, M.: DynNet: physics-based neural architecture design for nonlinear structural response modeling and prediction. Eng. Struct. **229**, 111582 (2021). https://doi.org/10.1016/J.ENGSTRUCT.2020.111582
17. Zhou, J.M., Dong, L., Guan, W., Yan, J.: Impact load identification of nonlinear structures using deep recurrent neural network. Mech. Syst. Signal Process. **133**, 106292 (2019). https://doi.org/10.1016/J.YMSSP.2019.106292
18. Wang, L., Liu, Y., Gu, K., Wu, T.: A radial basis function artificial neural network (RBF ANN) based method for uncertain distributed force reconstruction considering signal noises and material dispersion. Comput. Meth. Appl. Mech. Eng. **364**, 112954 (2020). https://doi.org/10.1016/J.CMA.2020.112954
19. Sadeghi Eshkevari, S., Cronin, L., Sadeghi Eshkevari, S., Pakzad, S.N.: Input estimation of nonlinear systems using probabilistic neural network. Mech. Syst. Signal Process. **166**, 108368 (2022). https://doi.org/10.1016/J.YMSSP.2021.108368
20. Bishop, C.M.: Mixture density networks (1994)
21. Coleman, T.F., Li, Y.: An interior trust region approach for nonlinear minimization subject to bounds. SIAM J. Optim. **6**(2), 418–445 (2006). https://doi.org/10.1137/0806023
22. Hancock, J., et al.: An improved method of matching response spectra of recorded earthquake ground motion using wavelets. J. Earthq. Eng. **10** (2006). https://doi.org/10.1080/13632460609350629
23. Sundararajan, M., Taly, A., Yan, Q.: Axiomatic attribution for deep networks. PMLR, pp 3319–3328, 2017. Accessed 15 Oct 2021 [Online]. Available: https://proceedings.mlr.press/v70/sundararajan17a.html

Chapter 15
Simulation-Based Damage Detection for Composite Structures with Machine Learning Techniques

Alexandre Lang, André Tavares, Emilio Di Lorenzo, Bram Cornelis, Bart Peeters, Wim Desmet, and Konstantinos Gryllias

Abstract With the use of composite materials rising steadily in major industries such as the automotive and aerospace, a strong effort is put into structural health monitoring (SHM) and damage detection methods for these components. Nondestructive testing (NDT) techniques such as laser Doppler vibrometry (LDV) provide a valuable experimental setting for making measurements with dense grids of points without mass loading the structure. The use of machine learning (ML) and deep learning (DL) techniques for the subsequent classification of defects has the potential of creating a reliable and automated framework for damage detection. In this work, different ML and DL approaches were assessed for the task of detecting defects on a carbon fiber plate by using frequency response data. The approaches were enriched by considering simulated datasets (created with Finite Element Analysis) in a transfer learning framework. Simulation data is easier to generate than experimental, meaning any added value provided with simulation data is advantageous. A description of the obtained results of damage detection is presented, along with a comparative overview of the different techniques.

Keywords Damage detection · Composites · Simulation · NDT · Laser Doppler vibrometry · Machine learning

15.1 Introduction

Composite materials are composed of two or more distinct phases (matrix and fiber), whose overall properties are superior to those of the individual components. They are characterized by achieving high strength-to-weight ratio, high tensile strength at elevated temperatures, and good fatigue resistance, associated with a weight reduction compared to some of the individual components. A steeper trend has been recently experienced in the development and application of composite materials. They have been introduced in almost every industry in one way or another, but their use is most prominently observed in the automotive [1, 2] and aerospace industries [3, 4]. In this sense, many studies are developed for innovative ways of monitoring and detecting damages on these materials.

Within the aim of damage detection, nondestructive testing (NDT) techniques provide a wide range of analysis methods for inspecting materials without damaging or debilitating their further use. There are particularly advanced NDT techniques like laser Doppler vibrometry (LDV), which allows to measure high-frequency response of structures with dense

A. Lang
Siemens Industry Software NV, Leuven, Belgium

Department of Computer Science, KU Leuven, Leuven, Belgium

A. Tavares (✉)
Siemens Industry Software NV, Leuven, Belgium

Department of Mechanical Engineering, KU Leuven, Leuven, Belgium
e-mail: tavares.andre@siemens.com

E. Di Lorenzo · B. Cornelis · B. Peeters
Siemens Industry Software NV, Leuven, Belgium
e-mail: emilio.dilorenzo@siemens.com; bram.cornelis@siemens.com; bart.peeters@siemens.com

W. Desmet · K. Gryllias
Department of Mechanical Engineering, KU Leuven, Leuven, Belgium

Dynamics of Mechanical and Mechatronic Systems, Flanders Make, Lommel, Belgium
e-mail: wim.desmet@kuleuven.be; konstantinos.gryllias@kuleuven.be

R. Madarshahian, F. Hemez (eds.), *Data Science in Engineering, Volume 9*, Conference Proceedings of the Society for Experimental Mechanics Series, https://doi.org/10.1007/978-3-031-04122-8_15

measurement grids of points, without mass loading the structure being measured. In a high-frequency regime, the local defect resonance (LDR) concept states that the localized resonance activation of defects is achieved, making them easier to detect. This concept is based on the fact that the inclusion of a defect leads to a decrease in stiffness for a certain mass of the material in that area [5, 6].

Building on this context, machine learning (ML) techniques can be leveraged to develop automated methods capable of detecting damages with high accuracies. The application of ML techniques is often done after a previous step of feature engineering, in order to extract key information from raw data. Many experts use theoretical knowledge on the subject to select the best features to feed to the data-driven models for damage detection [7]. Alternatively, deep learning (DL) techniques can sometimes be leveraged in end-to-end approaches, where algorithms are in its turn employed directly on raw data. Such approaches can be found applied to damage detection, for example, on composite materials [7, 8], beam structures [9], rotating machinery [10, 11], electrocardiogram classification [12], and many more.

In this paper, damage detection for composite structures is studied, starting by performing laser Doppler vibrometry (LDV) measurements to a carbon fiber-reinforced polymer (CFRP) specimen. At a next stage, ML and DL algorithms are applied to create three methodologies for damage detection, namely, the Convolutional Neural Network (CNN), the Anomaly Detection Autoencoder (ADAE), and the Autonomous Anomaly Detection (AAD). The use of simulation data is leveraged within a context of transfer learning (TL) to increase the effectiveness of the CNN. A description of the different steps within these three methodologies and a comparative overview are presented.

This paper is organized as follows. In Sect. 15.2, the algorithms used in this work are presented. In Sect. 15.3, the measurement campaign is described. The Sect. 15.4 is dedicated to presenting the different methodologies for damage detection, along with the different results. Finally, the key conclusions are described in Sect. 15.5.

15.2 Deep Learning

Deep learning (DL) is a subset of ML techniques highly focused on artificial neural networks, with a degree of increased complexity considering application and ability to represent complex nonlinear functions. In this work, both convolutional neural networks and autoencoders were used.

15.2.1 Convolutional Neural Networks

Different from the more general matrix multiplication applied in fully connected neural networks, Convolutional Neural Networks (CNN) use the convolution mathematical operator for operations between layers of the algorithm. The architecture of a CNN is structured as a sequence of stages. In a first stage, two main types of layer operations are employed for feature extraction: the convolutional and pooling operations. In a convolutional layer, a filter or kernel is convolutionally multiplied across the input, generating a feature map. In a trained CNN, the feature maps extract key characteristics from the input. The pooling layer, usually applied after a convolutional layer, performs subsampling with a certain operation, like calculating the maximum or average of all values within a pooling layer's kernel. In a second stage, fully connected layers are employed for feature classification. The output of these layers will be adapted to the problem to which the CNN is being applied, whether predicting a label on a discrete classification problem or predicting a quantity on a continuous regression problem. This described dynamic of a neural network is represented in Fig. 15.1.

15.2.2 Autoencoders

An autoencoder is a type of artificial neural network used to learn efficient coding of unlabeled data. By attempting to regenerate the input from the encoding, the encoding is validated and refined, learning a lower dimensional representation of this input data. Autoencoders are composed of two sections: an encoder, which maps the input into a lower dimensional code, and a decoder, which maps the code back, reconstructing the input. In this encoding operation, the most significant features are learned and extracted from the input. Its architecture contains an input layer and output layer of the same size, connected by one or more hidden layers, similar to single-layer perceptrons, as exemplified in Fig. 15.2.

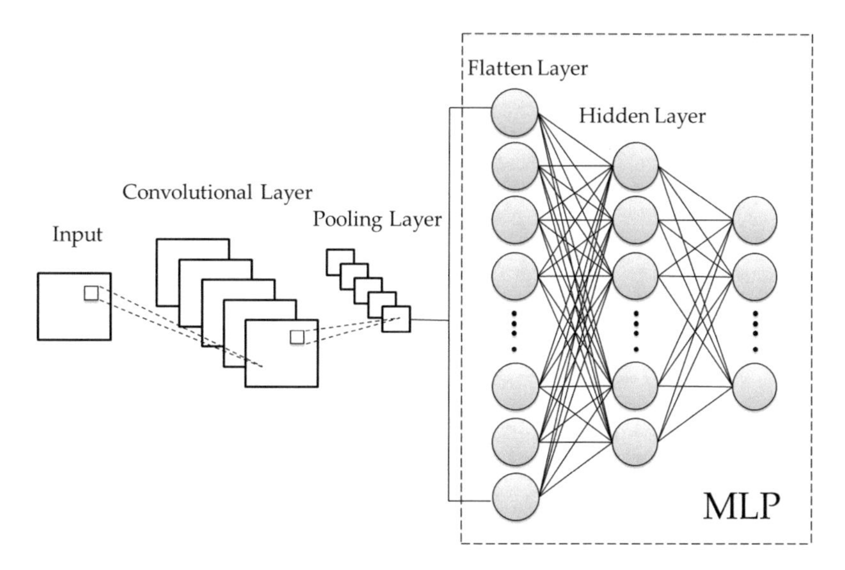

Fig. 15.1 Representation of the architecture of a Convolutional Neural Network (CNN)

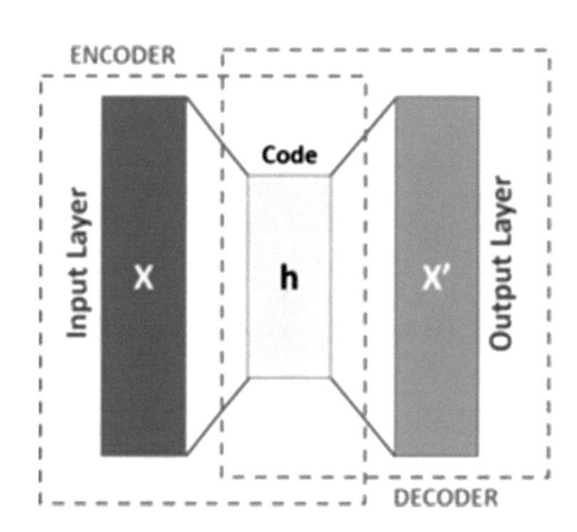

Fig. 15.2 Representation of the architecture of an autoencoder

15.3 Experimental Setup

A series of measurements have been performed using an Optomet SWIR laser Doppler vibrometer (LDV). This contactless measurement technique makes use of the Doppler effect associated with the scattering of a laser beam upon reflecting on the surface of a vibrating structure. It is a fast and simple solution to measure the vibration velocity, eliminating the mass loading on a structure imposed by more traditional measurement instruments like accelerometers or strain gauges. The test specimen is a carbon fiber-reinforced polymer (CFRP) plate (5.43 mm thickness) with flat bottom hole (FBH) damages (Figs. 15.4 and 15.5). In terms of experimental setup, the LDV was placed vertically with respect to the plate's top (undamaged) surface, which was placed on top of foam. A piezoelectric (PZT) patch was glued on the bottom surface of the plate, in order to provide the excitation. According to the local defect resonance (LDR) concept, to excite the plate into resonance behavior dominated by defect vibration, the frequency bands used to excite the test specimens should go up to high frequencies (e.g., in the order of kHz). For this study, the plate was excited up to 80 kHz using a chirp signal provided by the LDV's signal generator. Moreover, this signal was amplified 50 times before reaching the PZT patch, using the amplifier Falco Systems WMA-300. Figure 15.3 pictures the experimental setup described in this paragraph.

The CFRP plate is constituted by 24 laminae with $[(45/0/-45/90)]_{3S}$ stacking configuration and contains a total of 12 defects of varying diameter and thickness. Two measurements on the CFRP are considered for this work, one with the excitation of the PZT patch being provided in the center of the plate and the other with the excitation being provided in the corner, as shown in Figs. 15.4 and 15.5. The dataset hereafter referred to as Plate 1 had the PZT excitation in the center of the plate, and the dataset Plate 2 had the excitation in the corner of the plate. By exciting the plate on different locations, the time-response of the structure on the same point in the plate is different in the two datasets, and the relative distance of the damages to the excitation source is also varying. These variances cause both datasets to have fundamental differences.

Fig. 15.3 LDV experimental setup

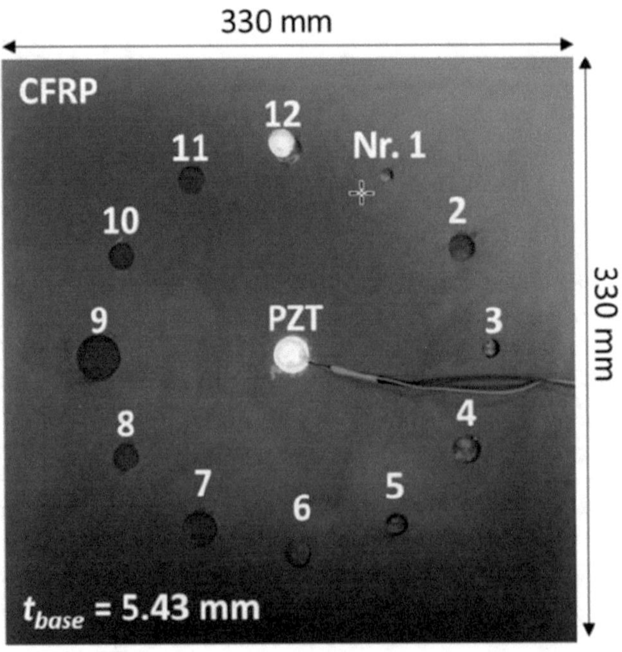

Fig. 15.4 Plate 1, with PZT patch in the center

Cross-validation techniques were considered for tuning the ML algorithms; therefore, both measurement datasets on the plate were horizontally split, to guarantee that the test set was located in a part of the plate which is independent from the training and validation sets. The horizontal split divided the Plate measurements into top halves (A) and bottom halves (B), as shown in Fig. 15.6. Plate 2B and Plate 1B were used for training and validation, respectively. Plate 1A was chosen as the

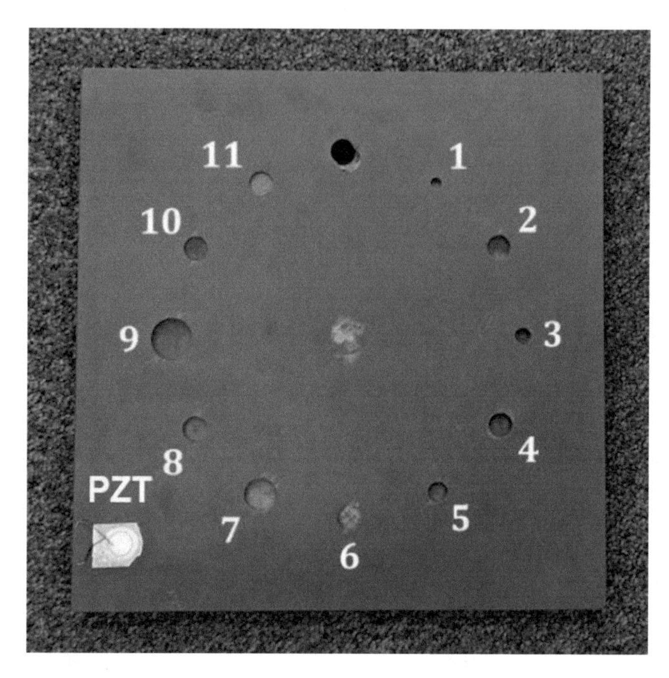

Fig. 15.5 Plate 2, with PZT patch in the corner

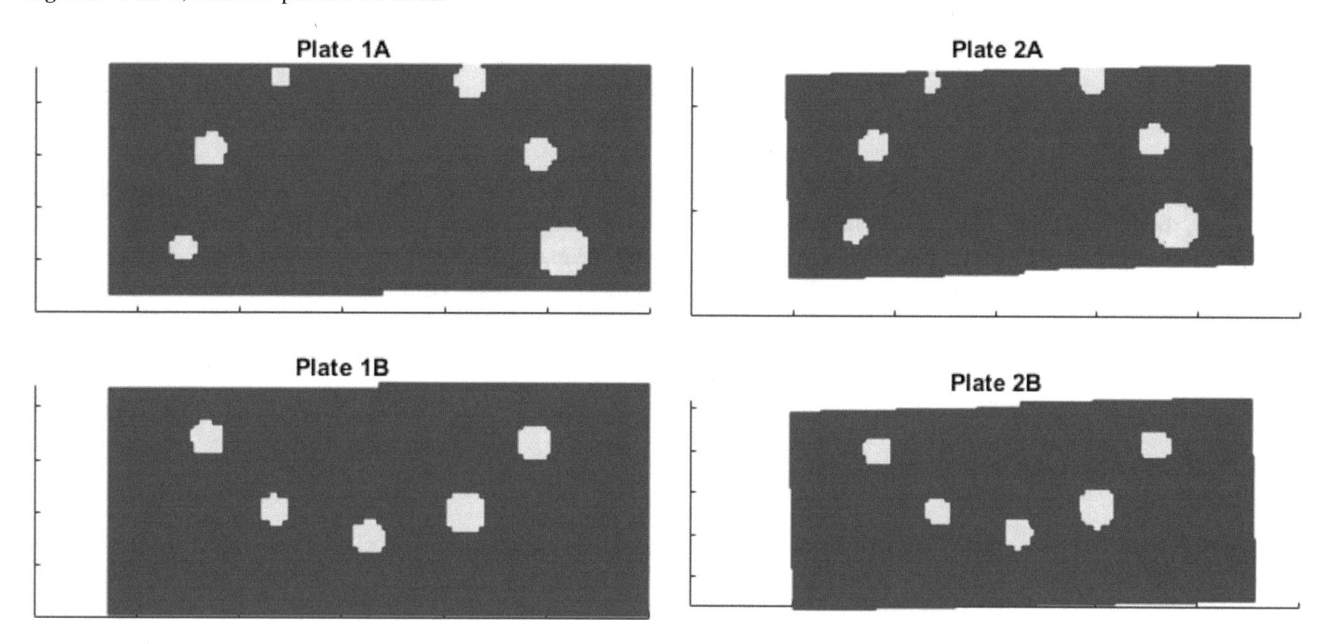

Fig. 15.6 Horizontal splitting of Plate 1 (left) and Plate 2 (right). The colors yellow and blue portray, respectively, the damaged and undamaged locations of the datasets

test set because this part of the specimen had the smallest defect (Nr. 1, according to Fig. 15.4). Plate 2A was not used, since it belongs to the same measurement as the training dataset (Plate 2B).

15.4 Analysis

Starting with vibration data acquired using the LDV with the experimental setup described in the previous section, several methodologies were developed for damage localization on the test specimen. These methodologies are composed of multiple stages: preprocessing, feature selection, machine learning techniques, and post-processing. A color-coded schematic is shown

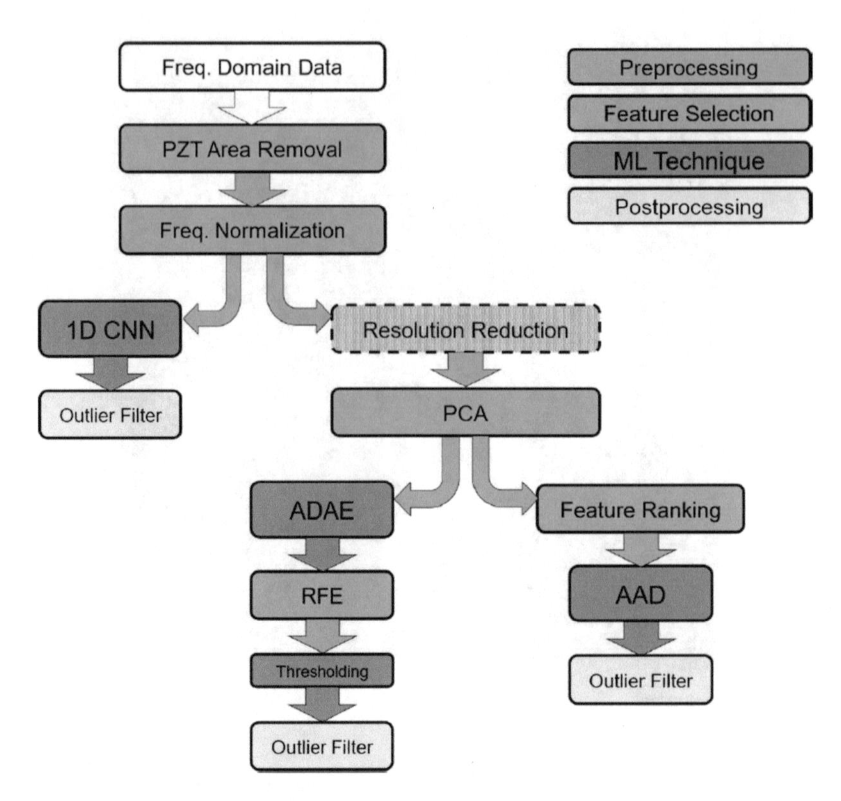

Fig. 15.7 Damage detection methodologies

in Fig. 15.7, with the sequential progression of the different stages per damage detection methodology. Considering ML techniques, three algorithms were used, being the basis of the three methodologies: the 1D Convolutional Neural Network (1D CNN), the Anomaly Detection Autoencoder (ADAE), and the Autonomous Anomaly Detection (AAD). Interestingly, each one of the algorithms corresponds to a different type of learner, this is, supervised learning, semi-supervised learning, and unsupervised learning, respectively. The development of these techniques was made in parallel and iteratively, in a way to also obtain an appreciative comparison of the different types of learners for the damage detection problem.

15.4.1 Preprocessing

As can be seen in Fig. 15.7, some preprocessing procedures were developed within the damage detection methodologies, namely, the PZT area removal, frequency normalization, resolution reduction, and principal component analysis (PCA).

Composite materials are neither homogeneous nor isotropic, which causes resistance to the propagation of energy through the material, and as a result it could be seen that the measurements near the PZT area had higher amplitudes than expected and were overall affected by this phenomenon. Figures 15.4 and 15.5 show the PZT patch attached, respectively, in the center and corner of the plate. The areas of the plate in these vicinities were therefore neglected from this damage detection study, since this data was unsuitable both for training and classification of the ML models.

Before applying a machine learning algorithm, the data from the measurements was normalized frequency-wise, to ensure that the variability inherent in having two different experimental measurements was reduced. Moreover, to reduce the number of parameters in the algorithms, it was also explored to reduce the resolution of the FRF data, by a factor of 4. Originally, each FRF contained 12,800 samples, and after resolution reduction each FRF contained 3200 samples.

For the ADAE and AAD algorithms, the data was projected to a lower number of dimensions through PCA. The eigenvectors used for projection came from the normalized Plate 1 dataset but were used for both the Plate 1 and Plate 2 datasets. Figure 15.8 shows the variance transmitted to the PCA projection according to the number of Principal Components (PC) used. It was interesting to observe that the variance of Plate 2 was quite successfully transmitted using the PCs from Plate 1, which shows that different experiments have common patterns and thus generalization is possible.

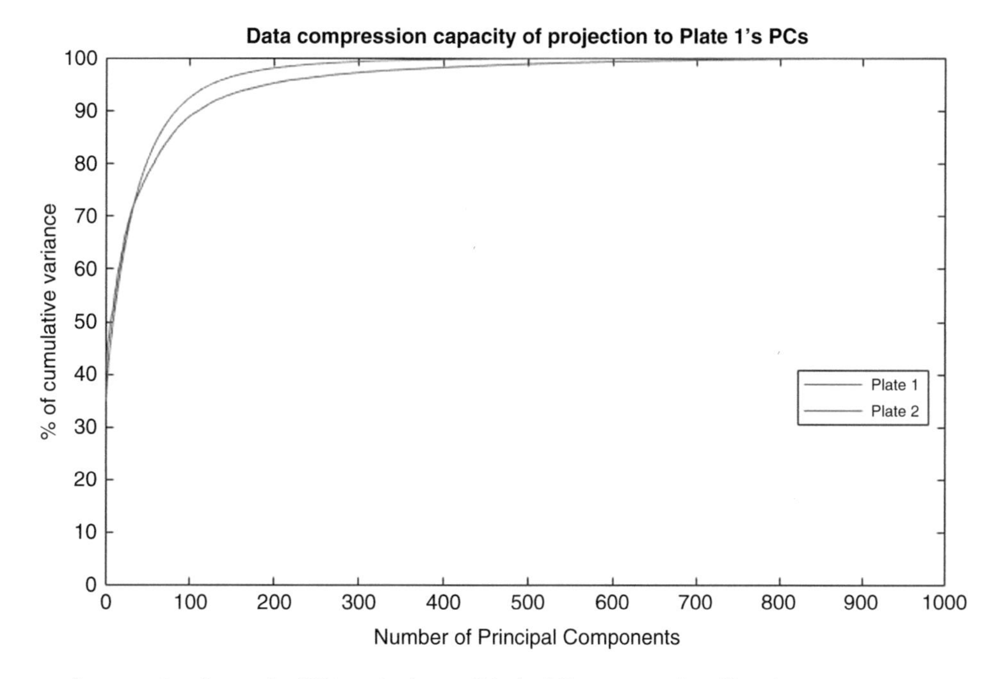

Fig. 15.8 Percentage of preserved variance after PCA projection on Principal Components from Plate 1

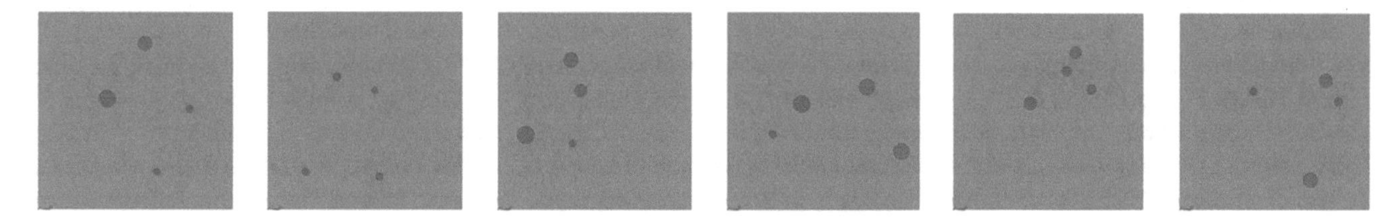

Fig. 15.9 Finite Element (FE) models of plates containing defects (red zones), used for training and validation of source models

15.4.2 Transfer Learning

Due to the large number of parameters associated with the CNN models, the available experimental data was rather limiting in its generalization capability, and Finite Element Analysis (FEA) was leveraged to create more data. Within this context, a transfer learning (TL) framework was explored with fine-tuning. That is, an initial CNN model was trained with abundant source domain data and then fine-tuned (retrained) with a more limited set of target domain data. In this application, the source domain data is the simulation one, obtained through Finite Element (FE) models which simulate a wide range of defect scenarios. Figure 15.9 shows the 6 FE models used to generate data for training the damage detection models. The target domain data is the experimental data obtained through LDV measurements.

15.4.3 Evaluation Metric

For evaluating the models, an evaluation metric like accuracy would not be suitable since non-defects correspond to more than 96% of the labels. The Intersection over Union (IoU) score, as defined by Eq. 15.1 (here a value of 1 corresponds to a defect, the index gt to the ground truth label, and p to the predicted label), was therefore selected as the evaluation metric, since it combines knowledge from metrics like true positives (TP) and false positives (FP). This score is commonly used for image segmentation, a task with a similar score context as the one of defect detection.

$$\mathrm{IoU} = \frac{\#\left(Y_{gt} == 1, Y_p == 1\right)}{\#\left(Y_{gt} == 1\right) \cup \left(Y_p == 1\right)} = \frac{\mathrm{TP} * \#\left(Y_{gt} == 1\right)}{\#\left(Y_{gt} == 1\right) + \mathrm{FP} * \#\left(Y_{gt} == 0\right)} \tag{15.1}$$

15.4.4 1D Convolutional Neural Network

The damage detection methodology using the 1D Convolutional Neural Network (1D CNN) involved the following steps: training and optimizing a model on simulation data (according to Sect. 15.4.2), fine-tuning the trained model on Plate 2B and optimizing it based on the validation dataset Plate 1B, and test on Plate 1A (according to Sect. 15.3. The optimization steps were performed using Bayesian optimization.

Bayesian optimization is a strategy for global optimization of black-box functions, which attempts to build a probabilistic model for the objective function and minimizes it by searching the optimal configuration of parameters through evidence it collects and an initial prior distribution [13, 14]. Therefore, both an objective function and the ranges for the different tunable parameters are given, for the optimizer to perform several iterations until finally giving an estimation of the best configuration it can find. First, a source model was optimized in terms of architecture and training hyperparameters for the simulation data. Having an optimized source model, the second Bayesian optimization step was to optimize the training hyperparameters for the fine-tuning step on experimental data. In the end, the best model had 9 convolution, ReLU (rectified linear unit), pooling (down sampling of order 2), batch normalization blocks (first convolutional layers having 16 filters, number which increased sequentially for the deeper layers), and 2 fully connected layers (with 400 neurons), trained with the ADAM solver [15].

Figure 15.10a shows the ground truth and Fig. 15.10b the best result obtained with the 1D CNN transfer learning model. The best IoU score obtained with this model was 0.516, and the overall IoU average for 15 trials was 0.398. Considering damage detection, 4 out of 6 flat bottom hole (FBH) defects were detected with some notion of their diameter. There is an additional classification of defected points on the top of the plate, which actually corresponds to a defected area of the plate, created by a manufacturing malfunction which was not included in the ground truth labels (can be seen in Figs. 15.4 and 15.5, near damage no. 12). The top two FBH defects, which are the most challenging, were not detected.

A TL framework may be affected if the target and source domains are not similar, in which case TL may end up actually deteriorating the final classification. In that case, it would be more beneficial not to use simulation data to pre-train a 1D CNN model and instead training only on the limited experimental target data, which is denominated here as learning from scratch (LFS). Figure 15.10c shows the results obtained from the model with a LFS methodology, whose performance is shown to obtain lower IoU score and overall noisier results, than the results obtained with the TL methodology in Fig. 15.10b. Although there is an inherent difference between the data obtained from FEA compared to the experimental data, the use of simulation data for pre-training models is shown to improve results. These can be associated to the fact that when a CNN model is trained on experimental data, it carries already pre-calculated parameters from the similar damage detection problem on simulation data, instead of initializing the training with random values for the algorithm's parameters. This can make both the model converge faster to a good set of parameters with the training and also transfer some knowledge from the similar damage detection problem on simulation data.

15.4.5 Anomaly Detection Autoencoder

The second approach developed for damage detection was the Anomaly Detection Autoencoder (ADAE), which is a semi-supervised learning technique. It consists of training an autoencoder with "healthy" data and, during testing, classifying an example as anomaly if its reconstruction error is above a certain threshold [16]. The Principal Components (PC) of the FRFs are the input to the autoencoder, which is built with a single hidden layer.

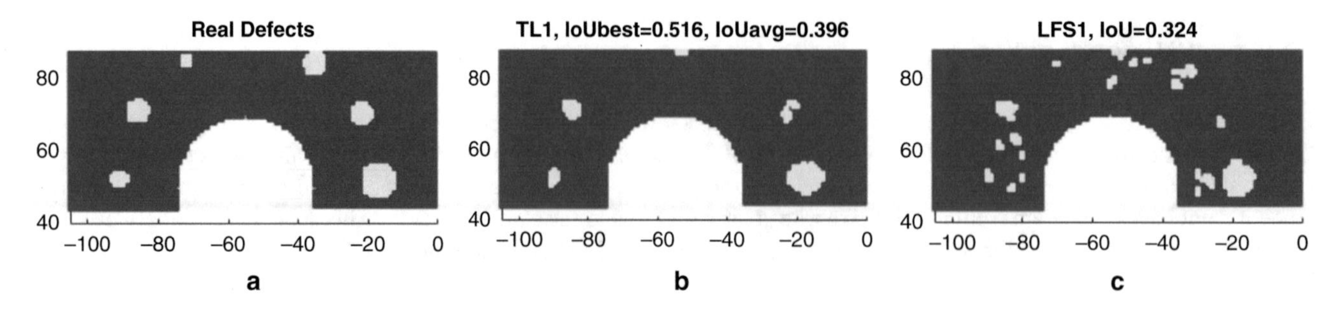

Fig. 15.10 Comparison of (**a**) ground-truth labels; with predictions of the 1D CNN for the: (**b**) Transfer Learning configuration, (**c**) Learning from Scratch configuration

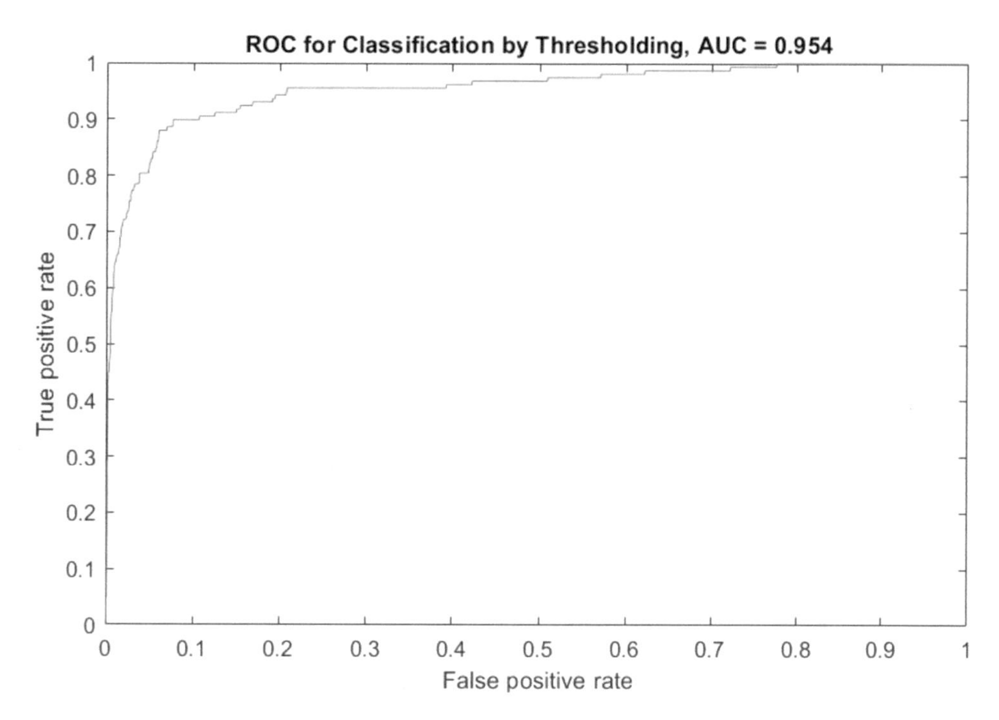

Fig. 15.11 ROC curve achieved with ADAE

Considering the thresholding, this value will act upon the reconstruction error, separating test examples to be classified as defects from the non-defects. When labels are available, for example, in training and/or validation sets, the threshold can be set in a way to maximize any given metric, which is here referred to as supervised thresholding. In this case, for each possible threshold, a corresponding value for the TP and FP rates can be plotted in a ROC curve, such as shown in Fig. 15.11. Moreover, following Eq. 15.1, since $\#(Y_{gt} = = 1)$ and $\#(Y_{gt} = = 0)$ are constant and the TP and FP are known, for each threshold value, IoU can be plotted as a function of threshold, as shown in Fig. 15.12.

The architecture of the ADAE and other hyperparameters, similar to the 1D CNN, were found through Bayesian optimization. The optimized model contains a single hidden layer with 150 neurons, training with 350 epochs and taking as input an overall of 300 Principal Components.

In order to improve the results of the autoencoder, the recursive feature elimination (RFE) method is employed to iteratively enhance the separability of the data by choosing features which were most important for the calculation of the reconstruction error. By applying the RFE on a trained autoencoder, the process is significantly faster since there is no need to train a new autoencoder each time. The RFE works by first calculating the squared error (SE) for each example and dimension, the sum of squared errors (SSE) for each example, and the IoU attained with all features. After this, the algorithm loops through each feature and checks the best attainable IoU when removing that feature's SE from the total SSE. The feature whose removal creates the most improvement is discarded, and the search restarts with the remaining features, until no feature removal improves the model. The RFE proved to be a good strategy to improve the IoU performance on the validation set.

Figure 15.13b shows the results obtained with the ADAE model. The detection of 4 out of 6 defects was achieved, plus the detection of the additional defect on the top center of the plot, which resulted from a manufacturing error. There was also a small misclassification located right to the defect on the bottom left corner of this Plate 1A. Figure 15.13c shows the results of the ADAE not using the step of recursive feature elimination (RFE) and therefore not using the validation set (Plate 1B). As can be seen, the IoU decreases considerably from 0.472 to 0.353 and thus demonstrates the added value of RFE and a thresholding tuned to RFE results, by considering the validation set.

15.4.6 Autonomous Anomaly Detection

The last technique to be tested was an unsupervised learning method called Autonomous Anomaly Detection (AAD), which requires no training data. The AAD is based on the work from Angelov and Gu [17, 18]. This algorithm finds anomalies on

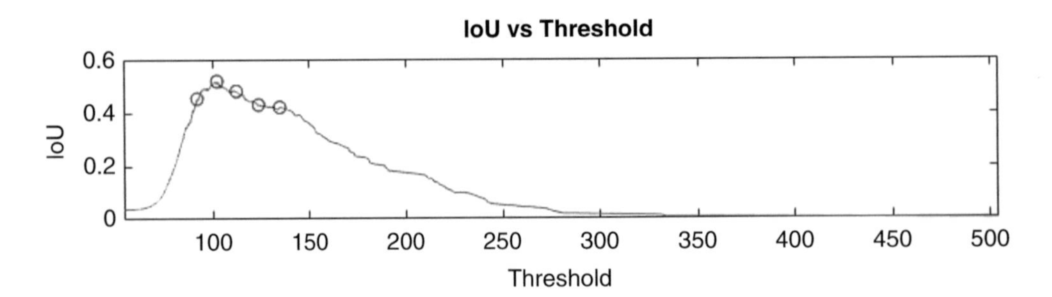

Fig. 15.12 IoU as a function of the threshold with peak values in red

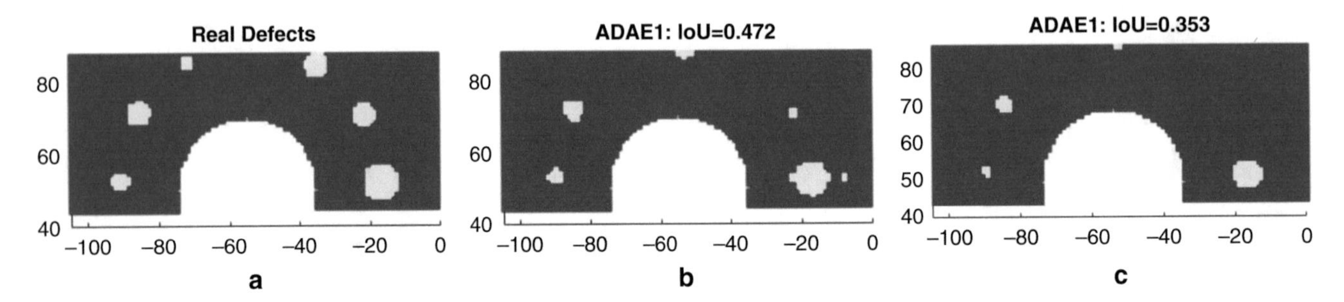

Fig. 15.13 Comparison of (**a**) ground-truth labels; with predictions of the ADAE: (**b**) with RFE, (**c**) without RFE

data, based on the empirical distribution of the data, and relies as little as possible on values and assumptions set by the user. This is an advantage in this damage detection context because different datasets can differ significantly. The first detection of potential anomalies by the AAD algorithm is based on the Chebyshev inequality expressing the probability of outlier points occurring within a set, shown in Eqs. 15.2 and 15.3.

$$P\left(\|\mu_K - x_i\|^2 \leq n^2\sigma_K^2\right) \geq 1 - \frac{1}{n^2} \tag{15.2}$$

$$P\left(\|\mu_K - x_i\|^2 > n^2\sigma_K^2\right) < \frac{1}{n^2};\ \text{for } i = 1, 2, \ldots, K \tag{15.3}$$

where μ_K is the mean of the dataset consisting of K elements; σ_K is the standard deviation of the dataset consisting of K elements; x_i is the element i of the dataset.

The value n can be chosen freely, but, in practice, $n=3$ is used most often, which is known as the 3σ principle. The algorithm flags as potential anomalies every point further than 3 standard deviations away from the mean, that is, it flags at most $1/n^2$ (1/9) points from the dataset. A distinction is made between global and local anomalies, as seen in Fig. 15.14: global anomalies, shown in red, can be found at first sight analyzing the data distribution; local anomalies, shown in green, do not have uncommon values for its variables but can be found according to the data density in their locations; collective anomalies, shown in magenta, are comparable to a cluster of local anomalies but may not always be considered as actual anomalies.

For this AAD methodology, the validation set (Plate 1B) was used for feature selection, while the training set was not used since no training was needed. The Feature Ranking was employed, which is a supervised approach that ranks features based on chi-squared tests, determining the relative importance of each feature for the classification shown in the validation set labels [19]. A grid search was performed for the best configurations regarding the number of Principal Components to use as input.

Through Feature Ranking, 60 out of the main 300 PCs were selected as input to the AAD, obtaining in the end a good IoU result of 0.498, as shown in Fig. 15.15b. Overall, 5 out of 6 defects were detected, plus the defect on the top center of the plate, although some false positives were also present, near the locations of the bottom defects. The smallest defect on the top left part of the plate was detected by this algorithm, which did not happen for the 1D CNN and ADAE, although more false-positive defects were also detected in comparison to these other methods. Figure 15.15c shows the performance of this

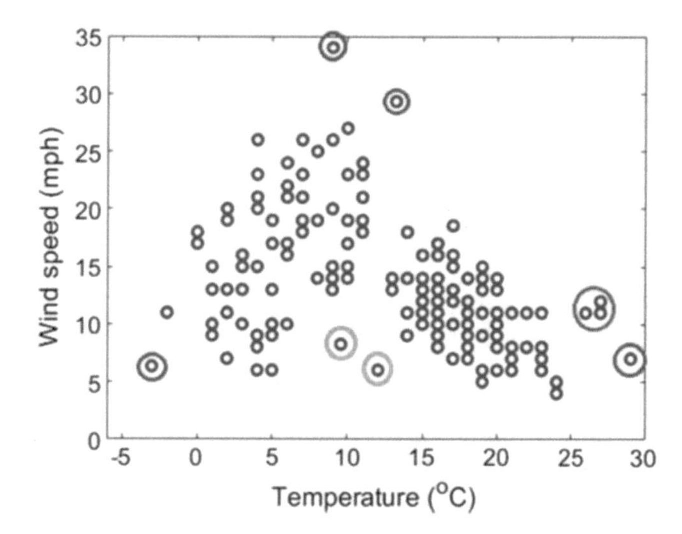

Fig. 15.14 Example of dataset containing global (red), local (green), and collective (magenta) anomalies. (Source: Adapted from Ref. [17])

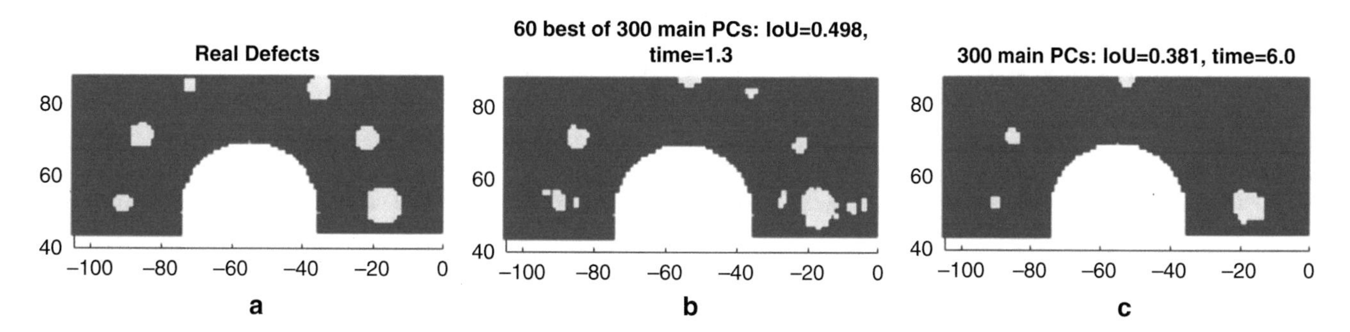

Fig. 15.15 Comparison of (**a**) ground-truth labels; with predictions of the AAD: (**b**) with Feature Ranking, (**c**) without Feature Ranking

algorithm without the Feature Ranking step, where the decrease in performance by the method can be observed. This can be attributed to not using the validation set to find the best Principal Components (PCs) for input, therefore possibly having as input certain PCs which are deteriorating the final classification result.

15.5 Conclusion

In this work, different ML and DL techniques were explored for the task of defect detection, namely, algorithms from the contexts of supervised, semi-supervised, and unsupervised learning. Each technique was also optimized in terms of hyperparameters, and simulation data was leveraged in a TL context for the 1D CNN. With the 1D CNN the highest IoU result was achieved, although there was some variability inherent in these results. The TL context proved to enhance the results obtained with this algorithm. However, the use of supervised learning requires labelled data on both healthy and damaged examples, a problem which affects less the other techniques, ADAE and AAD. The ADAE detected the same number of damages as the CNN, and RFE proved to improve its results. In comparison, one more defect was detected with the AAD technique, although some small misclassifications were observed in the final result. For this method, the feature selection also proved to improve the results. The detection of the smallest defect was never achieved, which may also be related to the fact that the measurement itself did not achieve a high enough frequency to activate it, according to the LDR concept. As next steps, higher frequency measurements will be performed, and the use of simulation data increased, in order to achieve higher accuracies. The work is also planned to be extended to different types of damages, where the trends of semi-supervised learning and unsupervised learning will be followed as well.

Acknowledgments The authors gratefully acknowledge SIM (Strategic Initiative Materials in Flanders) and VLAIO (Flemish government agency Flanders Innovation & Entrepreneurship) for their support of the ICON project DETECT-ION, which is part of the research program MacroModelMat (M3), coordinated by Siemens (Siemens Digital Industries Software, Belgium). André Tavares is supported by a VLAIO Baekeland PhD mandate [nr. HBC.2020.2300].

References

1. Rudd, C.D.: Composites for Automotive Applications. iSmithers Rapra Publishing (2000)
2. Ravishankar, B., Nayak, S.K., Kader, M.A.: Hybrid composites for automotive applications–a review. J. Reinf. Plast. Compos. **38**(18), 835–845 (2019)
3. Kumar, S., Reddy, K.M., Kumar, A., Devi, G.R.: Development and characterization of polymer–ceramic continuous fiber reinforced functionally graded composites for aerospace application. Aerosp. Sci. Technol. **1**(26), 185–191 (2013)
4. Cooper, T., Smiley, J., Porter, C., Precourt, C.: Global Fleet & MRO Market Forecast Commentary. Olyver Wyman (2018)
5. Solodov, I., Rahammer, M., Gulnizkij, N.: Highly-sensitive and frequency-selective imaging of defects via local resonance. In: Proceedings of European Conference on Non- destructive testing (ECNDT 2014), 2014
6. Solodov, I.: Resonant ultrasonic imaging of defects for advanced non-linear and thermosonic applications. International Journal of Microstructure and Materials Properties. **9**, 261–273 (2014)
7. Tavares, A., Di Lorenzo, E., Peeters, B., Coppotelli, G., Silvestre, N.: Damage detection in lightweight structures using artificial intelligence techniques. Exp. Tech. **45**, 389–410 (2021)
8. Aldrin, J.C., Forsyth, D.S.: Demonstration of using signal feature extraction and deep learning neural networks with ultrasonic data for detecting challenging discontinuities in composite panels. In: AIP conference proceedings, Melville, 2019
9. Abdeljaber, O., Avci, O., Kiranyaz, S., Gabbouj, M., Inman, D.J.: Real-time vibration-based structural damage detection using one-dimensional convolutional neural networks. J. Sound Vib. **388**, 154–170 (2017)
10. Chen, A., Mauricio, W.L., Gryllias, K.: A deep learning method for bearing fault diagnosis based on cyclic spectral coherence and convolutional neural networks. Mechanical Systems and Signal Processing. **140**, 106683 (2020)
11. Chen, K.G., Li, W.: Intelligent fault diagnosis for rotary machinery using transferable convolutional neural networks. IEEE Transactions on Industrial Informatics. **III**(29), 339–349 (2019)
12. Kiranyaz, S., Ince, T., Gabbouj, M.: Real-time patient-specific ECG classification by 1-D convolutional neural networks. IEEE Trans. Biomed. Eng. **63**(3), 664–675 (2015)
13. Mockus, J.: Bayesian Approach to Global Optimization: Theory and Applications. Springer (2012)
14. Bergstra, J., Bardenet, R., Bengio, Y., Kégl, B.: Algorithms for hyper-parameter optimization. In: Advances in neural information processing systems, no. 24, 2011
15. Kingma, D.P., Ba, J.: Adam: a method for stochastic optimization. arXiv preprint arXiv:1412.6980, 2014
16. Tsai, D.M., Jen, P.H.: Autoencoder-based anomaly detection for surface defect inspection. Advanced Engineering Informatics. **48**, 101272 (2021)
17. Angelov, P.P., Gu, X.: Empirical Approach to Machine Learning. Springer, Cham (2019)
18. Gu, X.: Self-Organising Transparent Learning System. Lancaster University (United Kingdom) (2018)
19. Thaseen, I.S., Kumar, C.A., Ahmad, A.: Integrated intrusion detection model using chi-square feature selection and ensemble of classifiers. Arab. J. Sci. Eng. **4**(44), 3357–3368 (2019)

Chapter 16
Synthesizing Dynamic Time-Series Data for Structures Under Shock Using Generative Adversarial Networks

Zhymir Thompson, Austin R. J. Downey, Jason D. Bakos, and Jie Wei

Abstract Validation of state observers for high-rate structural health monitoring requires the testing of state observers on a large library of pre-recorded signals, both uni- and multi-variate. However, experimental testing of high-value structures can be cost and time prohibitive. While finite element modeling can generate additional datasets, it lacks the fidelity to reproduce the non-stationarities present in the signal, particularly at the higher end of the digitized signal's frequency band. In this preliminary work, generative adversarial networks are investigated for the synthesis of uni- and multi-variate acceleration signals for an electronics package under shock. Generative adversarial networks are a class of deep learning approach that learns to generate new data that is statistically similar to the original data but not identical and thus augmenting the data diversity and balance. This chapter presents a methodology for synthesizing statistically indistinguishable time-series data for a structure under shock. Results show that generative adversarial networks are capable of producing material reminiscent of that obtained through experimental testing. The generated data is compared statistically to experimental data, and the accuracy, diversity, and limitations of the method are discussed.

Keywords Time-series · Machine learning · Adversarial network · Impact · High-rate dynamics

16.1 Introduction

A high-rate dynamic event is defined by its time scale of less than 100 ms and encountered high amplitude exceeding 100 g_n [1]. Hypersonic structures, automobiles during accidents, and active blast mitigation are examples of structures that undergo high-rate dynamic events. The goal of high-rate structural health monitoring (HRSHM) is to quickly assess and mitigate changes to a structure caused by high-rate dynamics [2]. One approach to achieve this goal is the development of state observers for tracking the state (i.e., damage) of the structure [3, 4]. The validation of observers for all possible states of high-rate systems is challenging; one reason for this is the inconsistent responses of the structure to these events as structural damage progresses through the life of the structure. As structures fail through a variety of modes, repetition of experiments results in tests with significant variations between tests. A sufficiently large dataset that covers the dynamics of the system would allow for the validation of state observers for high-rate systems.

High-rate events can cause plastic deformation to a structure, thereby requiring new structures for each successive experiment. The cost of the experiments themselves is another consideration as the cost of a test program scales with the price for the construction of the structure. Simulating high-rate events is a complex task that requires making various modeling assumptions that may miss key first order effects while ignoring various follow-on effects for the sake of stability. Additionally, the introduction of non-Gaussian sensor noise to the system requires the making of various assumptions and it is computationally expensive to rerun tests for each considered noise case. These technical and financial costs limit the potential

Z. Thompson · A. R. J. Downey (✉)
Department of Mechanical Engineering, University of South Carolina, Columbia, SC, USA
e-mail: zhymir@email.sc.edu; austindowney@sc.edu

J. D. Bakos
Department of Computer Science and Engineering, University of South Carolina, Columbia, SC, USA
e-mail: jbakos@cse.sc.edu

J. Wei
Department of Computer Science, The City College of New York, New York, NY, USA
e-mail: jwei@ccny.cuny.edu

R. Madarshahian, F. Hemez (eds.), *Data Science in Engineering, Volume 9*, Conference Proceedings of the Society for Experimental Mechanics Series, https://doi.org/10.1007/978-3-031-04122-8_16

size of the dataset. Utilizing machine learning for the development of high-rate structural health monitoring algorithms requires large amounts of data. Generating a large dataset of examples would require developing an extensive experimental and/or numerical testing campaign. There is a decent likelihood that there exist combinations of variables that are not present in the dataset but exist in real-world scenarios for a multivariate problem.

The data size and type limitation can be circumvented through generative models. The type of generative model chosen in this case was the Generative Adversarial Network (GAN). GAN is a deep learning algorithm for the synthesis of statistically indistinguishable data. Collecting real-world data is both expensive and time consuming, but GANs can mitigate this by producing statistically indistinguishable data based on a library of prerecorded events. In data collections, some events or classes are rare or hard to get, resulting in highly unbalanced dataset, which will cause great trouble for future effective training and analysis. GANs can be used to generate data items for these events or classes, thus mitigating the annoying unbalanced data problem. Furthermore, GANs can produce combinations of data not found in the original dataset. This means experimental data can be more focused on creating a representative sample, and minor variations can be artificially generated in the future. The purpose of the generative model is to augment the existing data to yield a diverse and balanced set of convincing results for use in the training or testing of state observers. GANs have been used to generate audio, radar, optical, and EEG signals [5–8]. As of the writing of this chapter, to the best knowledge of the authors, no work involving GANs to produce high-rate dynamic vibration signals has been reported. The contributions of this chapter include: (1) a GAN trained on high-rate dynamic vibration data, and (2) development of a multi-modal conditional GAN for multi-class interpolation.

16.2 Background

Generative Adversarial Networks are a type of deep learning technique capable of producing realistic data similar to, but not the same as, the given set of input data. They accomplish this through a minimax adversarial game in which a discriminator and generator compete [10]. The discriminator attempts to tell apart the real examples and those "fake" ones produced by the generator. The generator attempts to produce new data out of random noises dictated by the latent space that appears authentic enough to fool the discriminator. This relationship is described with the equation:

$$\min_G \max_D V(D, G) = \mathbb{E}_{x\sim p_{\text{data}}(x)}[\log D(x)] + \mathbb{E}_{z\sim p_z(z)}[\log(1 - D(G(z)))]. \tag{16.1}$$

Throughout training, the discriminator tries to maximize the function and the generator attempts to minimize it until they reach a balance. Wasserstein GAN (WGAN) is an alternative approach to the typical GAN setup [11]. WGAN transforms the objective function for the discriminator and generator. Rather than maximizing and minimizing the log likelihood that a given generated example is not genuine, the discriminator, now called a critic, attempts to minimize an approximation of the Earth Mover Distance while the generator attempts to maximize it [11].

$$f : X \to \mathbb{R} = \max_{\|f\|_L \le 1} \mathbb{E}_{x\sim\mathbb{P}_r}[f(x)] - \mathbb{E}_{x\sim\mathbb{P}_\theta}[f(x)]. \tag{16.2}$$

The change to the objective function converts the gradient slope for backpropagation from a sigmoidal to a linear shape. This change helps prevent convergence failure by vanishing gradients and simultaneously reduces the likelihood of the generative model undergoing mode collapse.

The conditional WGAN (CWGAN) is merely a combination of the conditional GAN (CGAN) and the WGAN where a CGAN is a GAN that takes an additional input (i.e., a class label) for use in the discriminator and generator.

16.3 Methodology

The model architecture used for training was a conditional WGAN composed of CNNs. The discriminator had 8 convolutional layers followed by a single dense layer for reduction. Activation functions were all leaky ReLU and were only applied to the convolutional layers. The conditional data was applied just before the final layer. The generator had an initial dense layer with 7 convolutional layers after. The dense layer had no activation function, and all except for the last two convolutional layers had ReLU activation. The last two had a cosine activation and a tanh activation, respectively[5]. Both

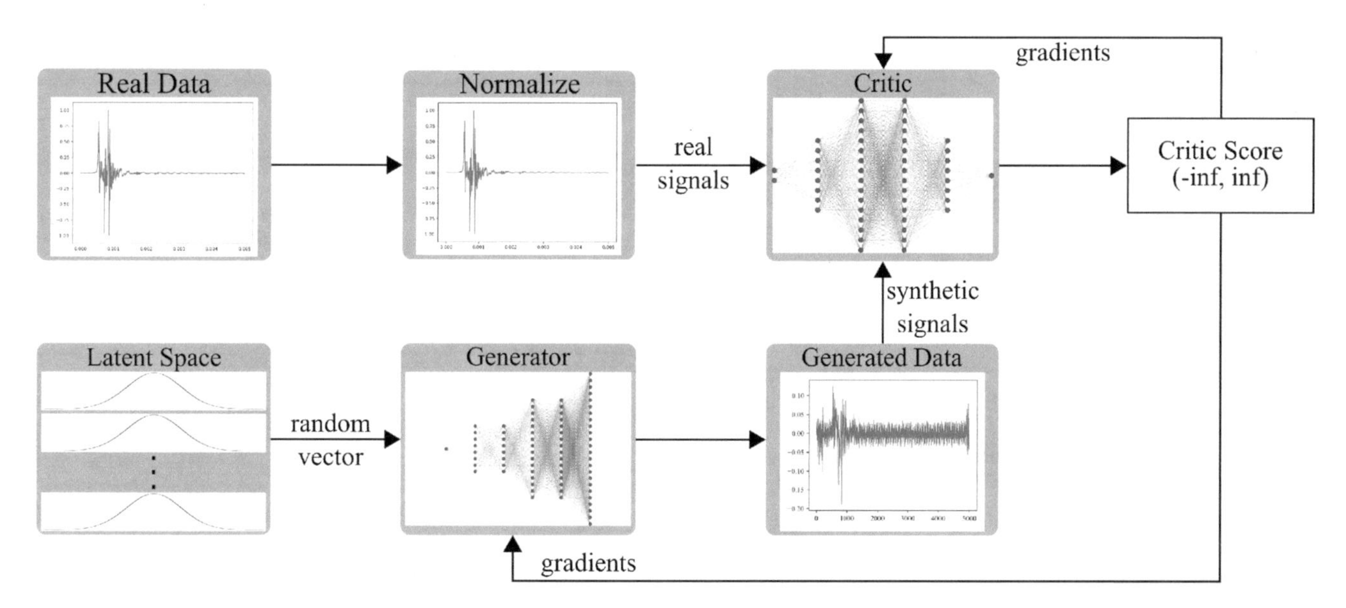

Fig. 16.1 An electronics package developed by Dodson et al. [9] where accelerometers were mounted on each unpopulated PCB board housed in axial fixture and held in place with a single lock ring torqued at 25 ft-lbs; the right-hand side of the figure shows examples of the measured accelerations for the four PCBs

used the Adam optimizer, but the discriminator had a lower learning rate of 0.0001 while the generator had a learning rate of 0.0002.

For model training, the event data was collected from multiple sources after experimentation was complete [9]. The experimental setup is shown in Fig. 16.1. The accelerometers, attached to PCBs, were placed in an axial test fixture for the duration of the experiment. The test fixture was held steadfast to a plate in the shock test system. Each experiment consisted of the system inducing a shock loading to the PCBs that was measured as acceleration. Data was digitized at 1 MS/s for every 5-ms experiment.

All signals were normalized to be on a scale of $[-1, 1]$. Due to the limited data size provided [9], the normalized signals were repeated so that there were about 50,000 samples in the training dataset. The generative model was given a 128-point vector for its latent dimension. The models were trained on batches of 16 samples where each batch passed through the critic 5 times, followed by the generator for one iteration. W1 regularization was applied to the critic to stabilize critic scores. Early stopping based on an FFT metric (though the critic score for generator could work) was applied, so training took less than 20 epochs to run.

Figure 16.2 shows the general flow of the training process. The real data used for training was first normalized to have the same shape but a range between $[-1, 1]$, and the output of the generator was set to have the same range. The condensed range reduced convergence time without significantly hindering the generator's production power since the generated data could be easily scaled back up to the real range in practice. An array of points drawn from a Gaussian distribution were passed to the generator, and the critic loss was calculated on the output from the generator and a batch of training data. The critic was further regularized using w1 regularization wherein the training data regularized the critic[12]. This looped for five iterations before continuing on to train the generator. The generated batch created previously was passed to the newly trained critic since the generator's predictions were deterministic and had not been trained since the generated data was produced. The generator trained on the critic's score of the data. This is a standard training loop for a GAN where the critic can be trained for longer since it avoids the vanishing gradient problem.

16.4 Analysis

Evaluating the efficacy of GANs is notoriously difficult as GANs lack an objective function [13]. Furthermore, the existing models for evaluation are mostly used for synthetic image generation and rely on pre-trained models built on that assumption. To evaluate the results of our GAN, an approach using the signals' frequency domain was modified to find the best fit for some given set generated signals in comparison to some set of reference signals [14]. Specifically, the frequency domain for each reference signal and generated signal was computed using the Fast Fourier Transform (FFT). Each generated signal was

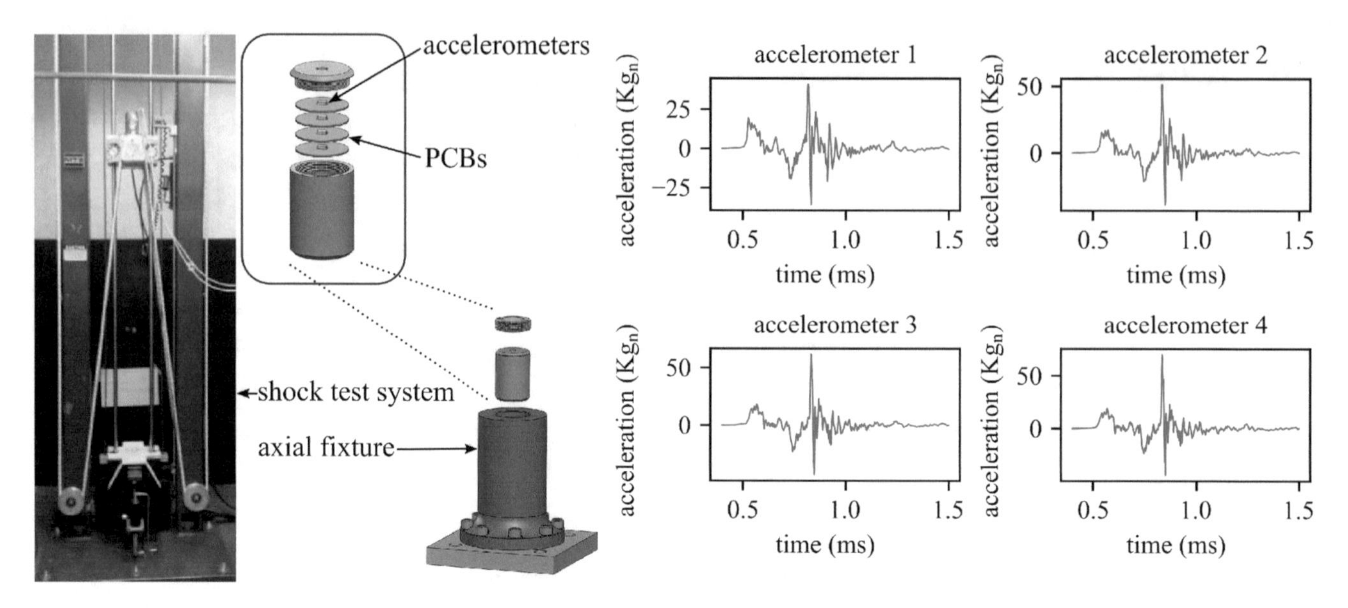

Fig. 16.2 A flowchart of training steps utilized for the proposed GAN methodology

Table 16.1 Scoring for generative model; the total is FFT score over all accelerometer data and each subsequent column is an FFT score for the respective accelerometer data

	Total	Accelerometer 1	Accelerometer 2	Accelerometer 3	Accelerometer 4
FFT score	1.496	3.161	3.449	4.017	1.987

compared to every reference signal, and the minimum mean squared error (MSE) was saved. Finally, the square root of the average of these differences was taken as the end score. The equation is summarized below:

$$f(x, y) = \frac{1}{m} \sum \min \left(\text{MSE}[\text{FFT}(x) - \text{FFT}(y)]\right), \tag{16.3}$$

where x is a batch of real samples, y is a batch of generated samples, and m is size of y. This score was found to correlate well with the convergence of the generator. While this score can observe accuracy for reproducing a signal, it fails to detect mode collapse, so all generated signals could be virtually identical. The scores in Table 16.1 were calculated with this scoring method.

Along with the FFT metric, samples were manually inspected to determine visual similarity. Manual inspection allowed for identifying potential mode collapse. Figure 16.3 shows the model's ability for generating examples from single accelerometers. Column (a) shows an arbitrary sample from each accelerometer of the training data, where row 1 corresponds to accelerometer 1 of the data. Column (b) consists of a single generated sample of the acceleration data for each of the four classes used.

The chosen samples provide a context for the generated data rather than a display of accuracy. The generated data now have a comparison for the general shape they were meant to learn. This model was able to produce examples from combinations of accelerometers as shown in Fig. 16.4.

Column (a) of Fig. 16.4 includes an arbitrary example from each accelerometer that the generative model was meant to combine into a single output. The corresponding output is shown in column (b). It is clear that the generative examples contain elements of the training data (oscillations centered around zero, low amplitude left tail, longer medium-low right tail, high amplitude center), so it is clear the model has found some relationship. Figure 16.4 excels at visualizing relationships, but in place of a multimodal mapping function, it details the relationship between each pairwise set of points in an example.

The plot layout for Fig. 16.5 is the same as in Fig. 16.3, but Fig. 16.5 contains the recurrence plots for the data rather than the data. These recurrence plots are generated by plotting the difference as a heat-map where points closer together are dimmer, of the acceleration for each given point in time against the acceleration of all other points in time. For example, each plot has a low-valued diagonal from the bottom-left to the top-right since in a pair of accelerations for any point in time with itself the distance would be zero. The important point to note is that similar plots have points with similar relationships, implying their frequency components are similar in the case of these waves.

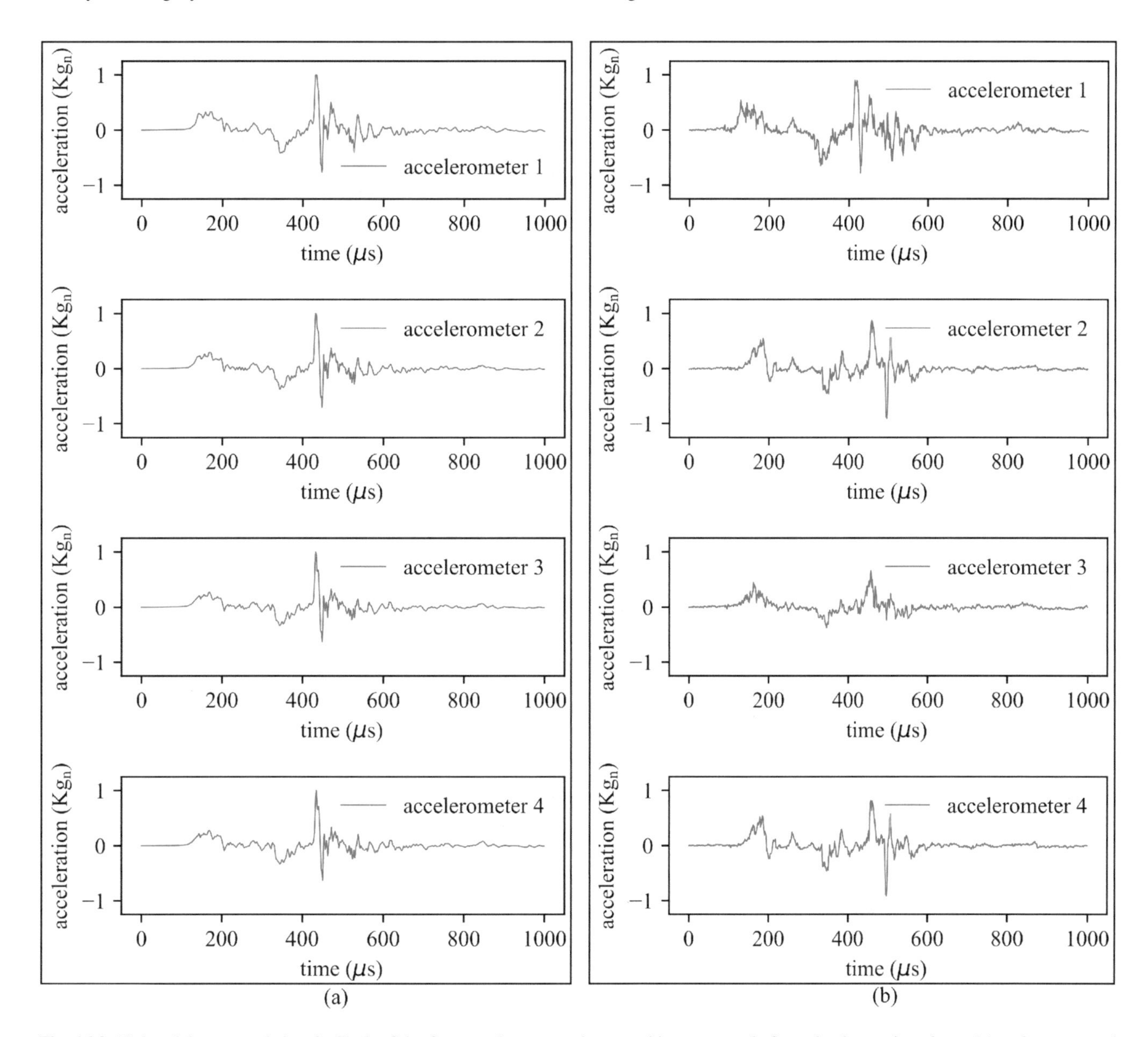

Fig. 16.3 Unimodal generated signals. Each of the four accelerometers has an arbitrary example from the dataset in column (**a**) and a generated example in column (**b**)

16.5 Conclusion

This chapter proposes a solution for the lack of consistency for high-rate events via GANs. Using a sample of experiments, the generator can find a mapping across the latent space. This allows for better consistency since a generator given the same input will always return the same output. Given the FFT scores and manual inspection, our model is capable of producing realistic synthetic signals as well as combinations of different signals. The variations between signals of the same class are minor, and variations between classes are not fully representative of the data they originated from. Mode collapse is a recurring problem for some model architecture instances with small batch sizes. The model architecture generated by this work did not show evidence of mode collapse. The results we obtained so far are encouraging. Different choices of latent space dimension, network architecture, dropout rates, optimization method, and learning rate combinations between the discriminator and generator can lead to vastly different results, and more investigations along these lines will be explored in the near future.

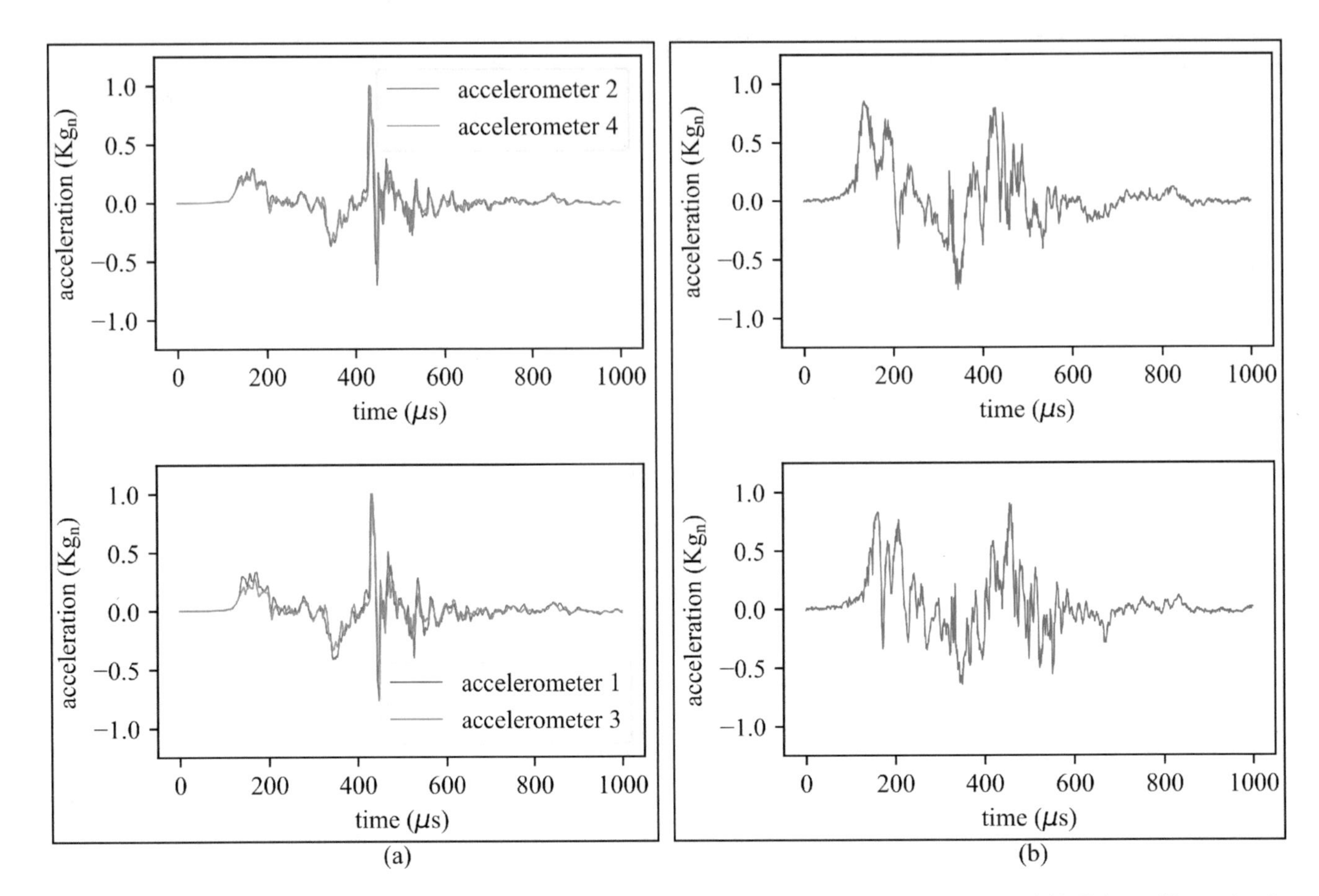

Fig. 16.4 Multimodal generated signals. Column (**a**) contains examples from the classes used for generation overlaid. Column (**b**) contains the corresponding generated examples

Acknowledgments This material is based upon work supported by the Air Force Office of Scientific Research (AFOSR) through award no. FA9550-21-1-0083 and no. FA9550-21-1-0082. This work is also partly supported by the National Science Foundation grant numbers 1850012, 1956071, and 1937535. The support of these agencies is gratefully acknowledged. Any opinions, findings, and conclusions, or recommendations expressed in this material are those of the authors and do not necessarily reflect the views of the National Science Foundation or the United States Air Force.

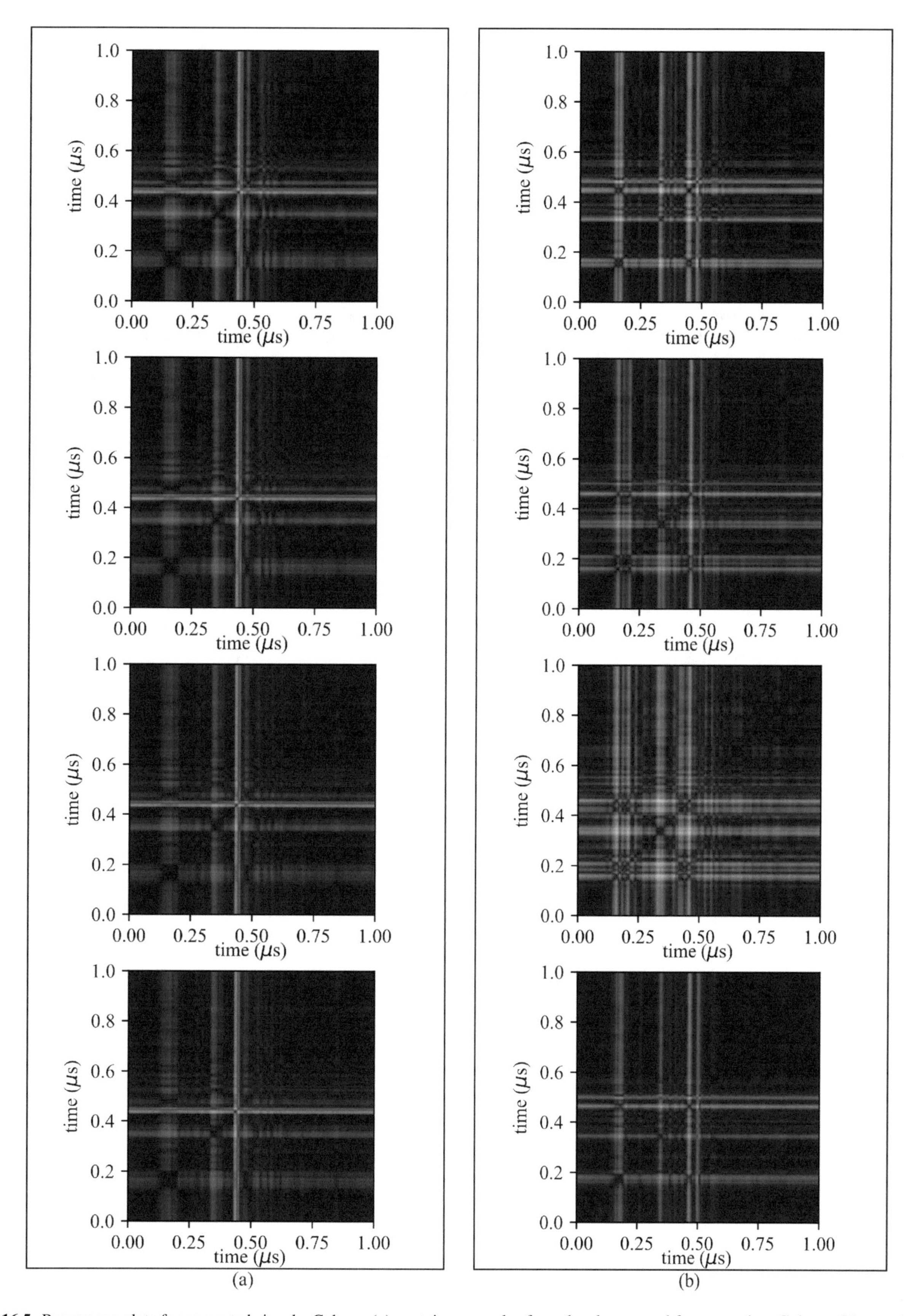

Fig. 16.5 Recurrence plots for generated signals. Column (**a**) contains examples from the classes used for generation. Column (**b**) contains the corresponding generated examples

References

1. Hong, J., Laflamme, S., Dodson, J., Joyce, B.: Introduction to state estimation of high-rate system dynamics. Sensors **18**(2), 217 (2018)
2. Dodson, J., Downey, A., Laflamme, S., Todd, M., Moura, A.G., Wang, Y., Mao, Z., Avitabile, P., Blasch, E.: High-rate structural health monitoring and prognostics: an overview. In: IMAC 39, Feb 2021
3. Hong, J., Laflamme, S., Cao, L., Dodson, J., Joyce, B.: Variable input observer for nonstationary high-rate dynamic systems. Neural Comput. Appl. **32**(9), 5015–5026 (2018)
4. Downey, A., Hong, J., Dodson, J., Carroll, M., Scheppegrell, J.: Millisecond model updating for structures experiencing unmodeled high-rate dynamic events. Mech. Syst. Signal Process. **138**, 106551 (2020)
5. Park, S., Han, D.K., Ko, H.: Sinusoidal wave generating network based on adversarial learning and its application: synthesizing frog sounds for data augmentation, Jan 2019
6. Donahue, C., McAuley, J., Puckette, M.: Adversarial audio synthesis, Feb 2018
7. Erol, B., Gurbuz, S.Z., Amin, M.G.: GAN-based synthetic radar micro-Doppler augmentations for improved human activity recognition, pp. 1–5. IEEE, Boston, MA (2019)
8. Hartmann, K.G., Schirrmeister, R.T., Ball, T.: EEG-GAN: generative adversarial networks for electroencephalograhic (EEG) brain signals, June 2018
9. Dodson, J., Hong, J., Beliveau, A.: Dataset 1 high rate drop tower data set, Dec 2019
10. Goodfellow, I.J., Pouget-Abadie, J., Mirza, M., Xu, B., Warde-Farley, D., Ozair, S., Courville, A., Bengio, Y.: Generative adversarial networks, June 2014
11. Arjovsky, M., Chintala, S., Bottou, L.: Wasserstein GAN, Jan 2017
12. Mescheder, L., Geiger, A., Nowozin, S.: Which training methods for GANs do actually converge? In: International Conference on Machine Learning 2018, Jan 2018
13. Salimans, T., Goodfellow, I., Zaremba, W., Cheung, V., Radford, A., Chen, X.: Improved techniques for training GANs. Adv. Neural Inf. Process. Syst. **29**, 2234–2242 (2016)
14. Lee, N.: How to measure the similarity between two signal? March 2015

Chapter 17
Multilayer Input Deep Learning Applied to Ultrasonic Wavefield Measurements

Cole N. Maxwell, Justin L. Dalton, Nicholas E. Dzomba, Erica M. Jacobson, Nikolaos Dervilis, and Adam J. Wachtor

Abstract Nondestructive evaluation (NDE) methods provide a way to monitor structures which can decrease maintenance/inspection costs, predict damage locations, and indicate the current state of structural health of a system. Acoustic steady-state excitation spatial spectroscopy (ASSESS) is an ultrasonic NDE technique that utilizes surface velocity response data collected by a laser Doppler vibrometer (LDV) to detect and characterize defects. Processing the resultant wavefield images produces a map of damage which is useful in manufacturing and structural health inspection of aerospace/civil structures.

Convolutional neural networks (CNN) were used to estimate plate defects directly from wavefield maps, producing pixel-wise thickness maps which are able to effectively characterize defects without producing boundary artifacts that occur with Fourier-based processing methods. This work utilized a diverse set of geometries and boundary conditions in its simulation-generated training set to ensure generalizability in the CNN. Using the U-net CNN architecture for image segmentation with a pretrained ResNet model as an encoder, a model was trained, validated, and tested which exhibits superior performance in speed and accuracy than traditional methods of defect characterization, while showing versatility over a wide range of plate shapes and boundary conditions.

Keywords Ultrasonic wavefield · Convolutional neural network · Acoustic steady-state excitation spatial spectroscopy

17.1 Introduction

Elastic waves can be used to detect damage within structures [1]. The study of how waves travel in a structure has allowed for identification of defects such as cracks, delamination, corrosion, and impact damage [2]. Lamb waves are a type of wave that propagate in plate-like structures, and a common way to generate them in structures is to use an excitation source such as a piezoelectric transducer attached to the structure [3]. Lamb waves are a valuable tool in nondestructive evaluation (NDE) as they can be used to detect and characterize defects in a structure. Reflections of the wave that occur at specific points in a

---

C. N. Maxwell
Engineering Institute, Los Alamos National Laboratory, Los Alamos, NM, USA

Department of Mechanical Engineering, Stanford University, Stanford, CA, USA
e-mail: maxwco@stanford.edu

J. L. Dalton
Engineering Institute, Los Alamos National Laboratory, Los Alamos, NM, USA

Department of Structural Engineering, University of California San Diego, La Jolla, CA, USA

N. E. Dzomba
Engineering Institute, Los Alamos National Laboratory, Los Alamos, NM, USA

Gianforte School of Computing & Department of Mathematical Sciences, Montana State University, Bozeman, MT, USA

E. M. Jacobson · A. J. Wachtor (✉)
Engineering Institute, Los Alamos National Laboratory, Los Alamos, NM, USA
e-mail: ejacobson@lanl.gov; ajw@lanl.gov

N. Dervilis
Department of Mechanical Engineering, The University of Sheffield, Sheffield, UK
e-mail: n.dervilis@sheffield.ac.uk

R. Madarshahian, F. Hemez (eds.), *Data Science in Engineering, Volume 9*, Conference Proceedings of the Society for Experimental Mechanics Series, https://doi.org/10.1007/978-3-031-04122-8_17

structure can be located through a variety of filtering methods [1]. Lamb waves travel through the structure in symmetric and antisymmetric modes. The key modes that are considered in NDE techniques are the first symmetric (S_0) and antisymmetric (A_0) modes [1]. The A_0 mode is a favorable choice for detection of damage in structures since it provides a high resolution due to its short wavelength [4].

A typical NDE system utilizing laser scanning ultrasonic detection methods consists of a laser Doppler vibrometer (LDV) and a device to excite the structure, such as a piezoelectric transducer or a laser, that is pulsed to create the ultrasonic waves [4, 5]. The inspection process is applied to a predetermined grid of locations at which the structure is excited. The resultant measurement matrix is three-dimensional (3D) – spatial data along two dimensions and temporal in the third. A method in which a steady-state excitation is applied to the structure was presented in [4]; the method is useful in that it increases the signal-to-noise ratio (SNR) in the measurement compared to those collected with transient excitation [5]. This method has been referred to as acoustic steady-state excitation spatial spectroscopy (ASSESS) or acoustic wavenumber spectroscopy. ASSESS is an NDE technique in which the structure is ultrasonically excited at steady-state; a pair of galvanometer mirrors steers the measurement location of the LDV, and the response of the structure's surface velocity is continuously recorded [6–9]. This method allows for high spatial resolution surface response measurements to be obtained orders of magnitude faster than transient measurements and enables the determination of spatial wavenumber of the response at each location in the spatial grid [5]. Thickness maps of the structure can then be determined by isolating a particular wave mode, determining the local wavenumber at each spatial grid location, and using Lamb wave dispersion curves to match wavenumber to thickness. Regions with differing wavenumbers, and therefore different thicknesses, can indicate a structural defect or change in the material when its designed thickness is uniform [5].

Steady-state excitation provides an advantage for NDE measurements since the surface response can be continuously and simultaneously collected as the structure is excited. Other ultrasonic NDE methods use transient wave measurements in which the transient wave must die out before another measurement is taken. This results in fewer overall measurement points compared to an ASSESS measurement for the same inspection duration. As such, steady-state measurements allow for full-field inspection of the entire structure. The speed of this inspection technique has also shown promise in performing in situ inspection of additively manufactured parts [10, 11].

While the LDV ideally scans perpendicular to the structure during and ASSESS measurement, scans of tilted surfaces can be manually corrected using an estimate of the perspective tilt angle [12]. Without this correction, the wavefield is artificially compressed from the perspective of the measurement system, making local wavenumber-based analyses inaccurate.

Recent work has implemented a convolutional neural network (CNN) to process ASSESS measurements, replacing original local wavenumber estimation methods [6]. Inspired by the receptive fields in the visual cortex, CNNs are a type of artificial neural network (ANN) frequently used for tasks involving visual imagery, such as image classification and image segmentation [13]. ANNs incorporate multiple layers with parameters (or weights) that are optimized through an iterative process called training. During training, a loss function evaluates the performance of the model over multiple training examples by comparing the output with the ground truth. By back-propagation, the parameters across the different layers of the models are optimized using stochastic gradient descent. CNNs differ from other ANNs in that at least one layer in a CNN performs convolution, the mathematical operation of combining two functions to create a new function [13].

The architecture of neural networks consist of an input layer, hidden computation layers, and an output layer. Hidden layers typically include weight layers which are updated during training and layers which perform specific mathematical operations like activation functions. ReLU (Rectified Linear Unit) is a common activation function which returns the input if the input is positive and zero otherwise. Such activation functions increase the expressiveness of deep learning models by introducing nonlinearities in the model [13]. All CNNs include a convolution layer, which contain a set of trainable filters (or kernels) that are convolved along the input volume. The dot product of the filters with the input produces an activation map which is able to capture features across the input space. The weights of the filters are shared, meaning that they do not change as the filter moves along the input. The weights are updated during training, and the filters are defined by hyperparameters, including the filter height, filter width, and the stride – which describes how the filter moves. Depending on the requirements of the model, multiple stacks of these layers can be sequenced to capture local and global information from the input data [13]. Fully connected layers take the output of the previous layer and perform an affine transform of the entire activation space using a matrix of trainable weights and a bias term, which is either fixed or trainable. The output of the fully connected layer produces a set of scores which can be used for classification, regression, or segmentation [13].

CNNs display superior performance compared to ANNs across a variety of tasks involving visual inputs. Although they are most frequently used in the two-dimensional (2D) space, recent work has also shown that they are useful for problems involving volumetric 3D data. There are several different approaches to working with 3D data, which often incorporate information about the geometry of the input. Work by Huang, et al. [14] projected 3D scans of pipe geometry into a 2D space as a preprocessing step before CNN classification of water leaks. In Velas, et al. [15], 3D LiDAR data in the Cartesian coordinate frame was directly incorporated as additional channels to the input for image segmentation of streets and roads.

Models which are known to be effective in 2D are commonly reconfigured for 3D data; for example, Çiçek, et al. [16] took the 2D U-net architecture, which is used for image segmentation, and increased the number of input channels for segmentation on 3D volumetric data to create the 3D-U-net model.

The U-Net CNN used by Eckels, et al. [3, 6] evaluated ASSESS data on a pixel-by-pixel basis to characterize defects of discrete thickness on an aluminum plate. The model was shown to yield fewer processing-related edge effects than the traditional ASSESS algorithm and had a reduced processing time. However, the CNN training assumed a flat inspection area that was perpendicular to the ASSESS measurement system, limiting the application space in which the model was effective. The primary objective of this work was to assess the performance of a CNN that incorporates both wavefield data and structural geometry information as inputs to the CNN for generating thickness maps for defect characterization on surfaces that are not perpendicular to the measurement system. By adding additional input channels for the imaginary component of the surface response and geometry information, the model is seen to be robust to wide range of geometric configurations.

17.2 Methodology

The working hypothesis driving this work is that incorporation of surface geometry information – the surface for which the response is measured with the LDV – will increase the generalizability of a CNN model and will enable thickness predictions for any measurement configuration with suitable SNR. Registering the ASSESS data to a surface geometry collected using a LiDAR, rangerfinder, or similar measurement alleviates the need to perform a perspective correction estimate on the wavefield before processing.

17.2.1 Simulation-Based Training Set

A total of eight datasets were simulated in ANSYS and were used to train the CNN [3, 6]. The datasets contained wavefield response measurements for flat, aluminum plates with a range of modeled defects of varying thicknesses: flat plates with no defects, a single defect, or multiple defects on the same plate. An excitation frequency of 80 kHz and plate area of 400 mm × 400 mm were constant for all simulations; transducer location, number of defects, defect locations, and defect thicknesses were varied within the datasets – Table 17.1.

17.2.2 Data Augmentation and CNN Input Preparation

The real and imaginary components of the response wavefields were interpolated onto a 1 mm^2/pixel grid. Several data augmentation procedures were utilized to increase the size and variety of the dataset without performing additional simulations. Data augmentation is a common practice in deep learning and produced new wavefields maps that had different characteristics but maintained the validity of the original data. The augmentations used included adding 5% Gaussian noise; rotations of 90°, 180°, and 270°; and vertical and horizontal transposes of the data. Additionally, the CNN was designed to have input resolution of 200 × 200 mm. As such, ten random crops of each wavefield were generated to reduce the 400 × 400 mm wavefields down to 200 × 200 mm. This procedure also increased the variety of boundary conditions in the dataset since cropped locations resulted in zero, one, or two free edges – Fig. 17.1.

The geometry mapping was performed by independent, random rotation of the wavefield segment in both the x-direction and y-direction between −45°and 45°. Using the general perspective transform, the distance was calculated from each pixel on the rotated plate to the focal center of the LDV, Fig. 17.2. This procedure was performed twice for each plate segment for two independent, random rotations of the plate, essentially doubling the training set. The input to the CNN was a three-channel tensor comprised of the real component of the response wavefield, imaginary component, and geometry mapping of the plate surface – Fig. 17.3.

Table 17.1 Simulation dataset information summary [3, 7]

	Dataset							
Parameter varied	1	2	3	4	5	6	7	8
# of defects	4	1	1	1	1	1	0	1
Transducer location	Center	One of nine potential locations	Center	Upper right quadrant	Upper right quadrant (4 options)	Lower left quadrant	Upper right quadrant (4 options)	Center
Plate thickness (mm)	10	10	10	10	10	2, 3, …, 9	1, 2, …, 10	10
Thickness at defect (mm)	2,4,6,8	2,4,6,8	2,4,6,8	1, 2, …, 9	1, 2, …, 9	1, 2, …, 8	n/a	1, 2, …, 9
Defect location	Corner and center edges	3 by 3 grid	Center	3 by 3 grid in upper right quadrant	center	center	n/a	top edge
Defect shape	Circle, square, rectangle	Circle, square, rectangle	circle, square	Circle, square	Circle	Square	n/a	Rectangle
Defect size (mm)	50 (diameter/edge), 10x50 (rectangle)	50 (diameter/edge), 10x50 (rectangle)	10–120 (diameter/edge)	50 (diameter/edge)	50 (diameter)	10, 30, …, 90	n/a	10, 20, …, 60
Dataset size	224	128	96	243	36	130	40	54

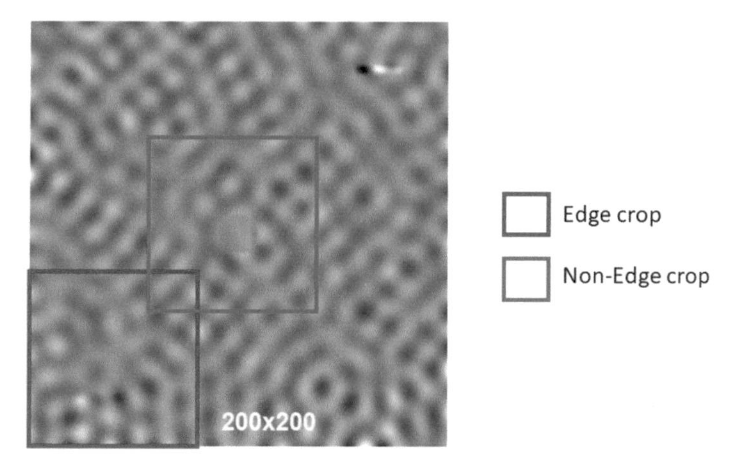

Fig. 17.1 Visualization of random cropping used in data augmentation which results in differing boundary conditions to the wavefield input

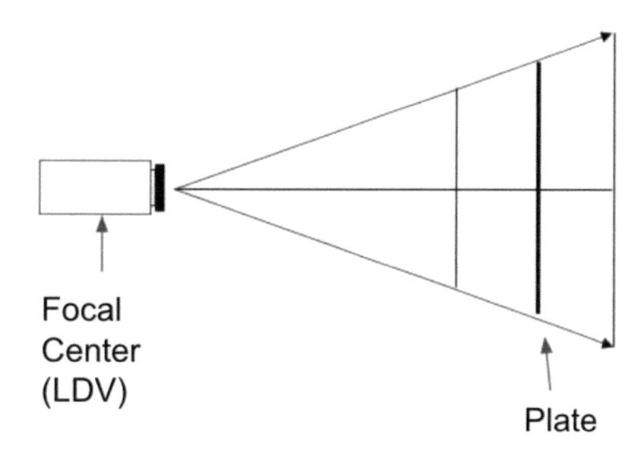

Fig. 17.2 A diagram showing the experimental setup of the LDV and plate

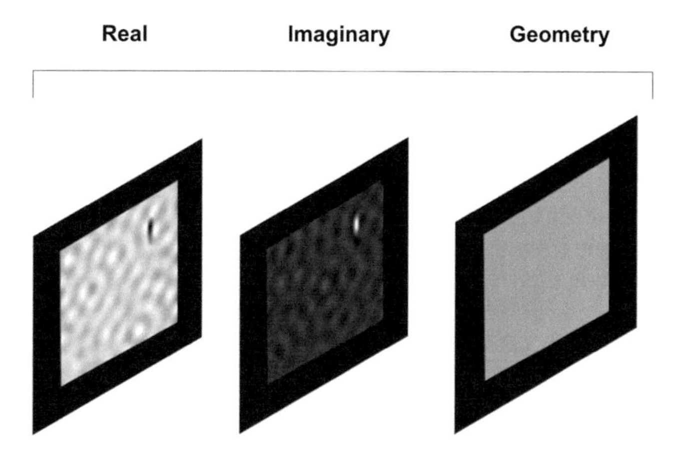

Fig. 17.3 A schematic of the CNN input, showing the real, imaginary of geometry channels

17.2.3 CNN Training

After data augmentation was performed and the data was prepared for input to the CNN, the simulation data was randomly split into a training set, validation set, and test set. The training set included 70% of samples, while the validation and test set each included 15% of samples. The training set, which contained 8064 samples, was used to update the model weights during training. After each epoch, the model was tested on the validation set of 1728 samples to avoid overtraining of the model on the training set. If model performance improved, the weights from the current iteration were saved. The 1728 sample test set was blind to the model during training and only used to provide an objective evaluation of model performance.

The CNN model was trained for 200 epochs. Model hyperparameters included a batch size of 32 and an initial learning rate of 0.0001. Adam – an adaptive optimizer – was used to tune the learning rate during training. Overall, the model contained 24,437,964 trainable parameters and 182 labels. A multi-label focal loss function was chosen for this model because the datasets were class imbalanced. First described in 2017, the focal loss function includes a modulating term that weights harder to learn examples greater than easy examples, which can overwhelm the training process when using a loss like the regular cross-entropy function which weights each label equally [17].

The specific CNN architecture used was a U-net encoder-decoder which incorporated the pretrained ResNet34 model as the encoder backbone. ResNet34 is a 34-layer CNN which has been previously trained on more than 100,000 images across 200 classes from the ImageNet Dataset [18]. This approach decreased the time needed for a model training and to improve overall model performance.

The intersection over union (IoU) score is a metric to determine the accuracy of the trained models. IoU tests for the overlap between the model's prediction and ground truth label. As every pixel in the input space is labeled, the IoU score is effectively a score of accuracy ranging from 0 to 1. The CNN model produced an IoU score of 0.9932 for the training set and an IoU score of 0.9748 for the validation set.

17.3 Results and Discussion

17.3.1 CNN Testing on Simulation Data

The accuracy of the model over every pixel in the test set was used to evaluate overall model performance. The IoU score for the entire test set was 0.945, meaning that 94.5% of pixels in the test set were labeled correctly. This score indicates that in general the model was able to detect thickness differences accurately. It should be noted that this score may be larger than expected due to a class imbalance in the training data for the CNN and the large number of 10 mm plates that it classified correctly which inflated the overall score.

A normalized confusion matrix was used to assess model performance by label – Fig. 17.4. The model performed best on the transducer, background, and 1 mm, 2 mm, and 10 mm thicknesses, correctly labeling these pixels with more than 95% accuracy. Model performance was lower for other labels, notably only correctly predicting 9 mm thicknesses with 32% accuracy; the confusion matrix shows that the model most frequently mislabeled these pixels as 10 mm. The datasets contained a large class imbalance, so increasing the training set size for those thicknesses which were seen little in the training phase would improve the values reported. Performing new simulations to achieve class imbalance was not permissible for this work, but will be done in future work.

Figures 17.5 and 17.6 show model predictions from the test set. As indicated by the confusion matrix in Fig. 17.4, it can be seen that the model performed better on test sets comprised only of the transducer, background, and 1 mm, 2 mm, and 10 mm thicknesses. For nominal plate thicknesses and defects that were between 3 and 9 mm, model predictions were less accurate and had more spatial variation in these thickness predictions than was contained in the ground truth. Despite these observations, thickness predictions were still generally in the correct range and differed from the ground truth predominantly by 1–2 mm. For example, the top row in Fig. 17.5 displays a 5 mm plate with predictions that were either 5 ± 1 mm across the plate. Thus, results from this CNN still contained relevant predictions, and by altering training schemes, performance can be improved. Additionally, defect regions were also identified, but the shape of the defect was not always captured well.

Figure 17.5 shows another set of model predictions for a set of plates each having a nominal thickness of 10 mm. Compared with the results in Fig. 17.4, nominal thickness predictions were much more accurate and the defect locations were increasingly distinct. In general, the transducer class was typically identified well by the model. The CNN model predictions show results without any indication of boundary effects, which is an improvement compared to Fourier-based processing.

17.3.2 CNN Testing on Experimental Data

To test the validity of the simulation-based CNN model on real-world measurements, experimental data was gathered on an aluminum plate of nominal 10 mm thickness with material removed from the opposite surface of that which measurements were performed – Figs. 17.7 and 17.8. The real component of the surface response for one of the measurements is in Fig.

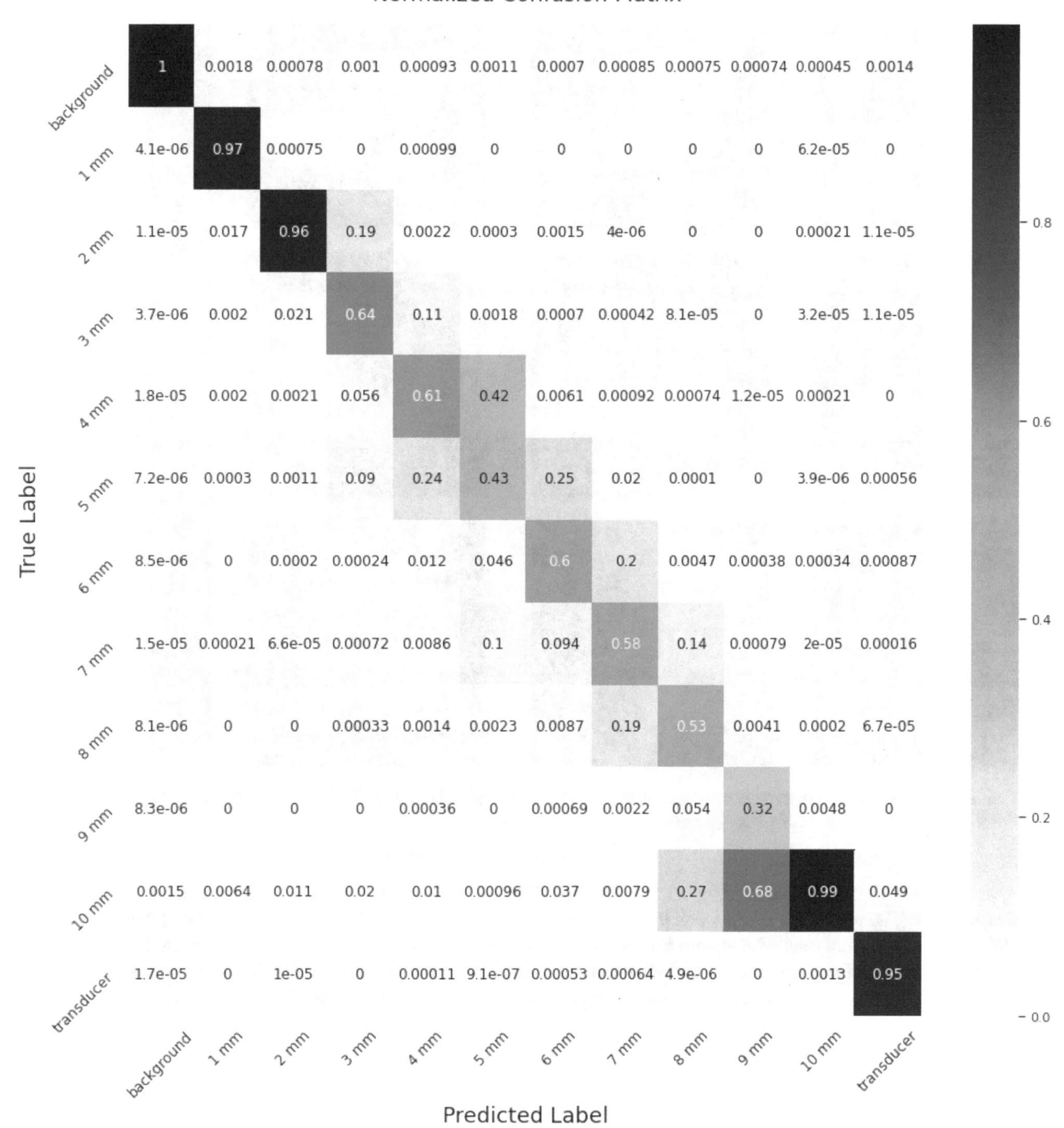

Fig. 17.4 Normalized confusion matrix for trained CNN model. The columns represent the true labels, while the rows represent predicted labels. Each cell was divided by the overall number of correct labels for the label indicated by the column

17.9. Data was collected in two configurations: (1) plate was perpendicular to the measurement system, Fig. 17.10, and (2) an angle of approximately 20° along the horizontal direction, Figs. 17.7 and 17.8.

A Velodyne Hi-Res Puck LiDAR was used to collect the geometry map, and a Polytec OFV-505 was used to record the surface response. The test specimens were located 2 meters from the measurement system and were excited at 80 kHz with an ultrasonic piezoelectric transducer. Data was collected at a sample rate of 2 MS/s, with anti-aliasing filters applied on the LDV controller at 100 Hz and 1.5 MHz. The LDV sensitivity was 1 m/s/V for the plate measurements. Each specimen

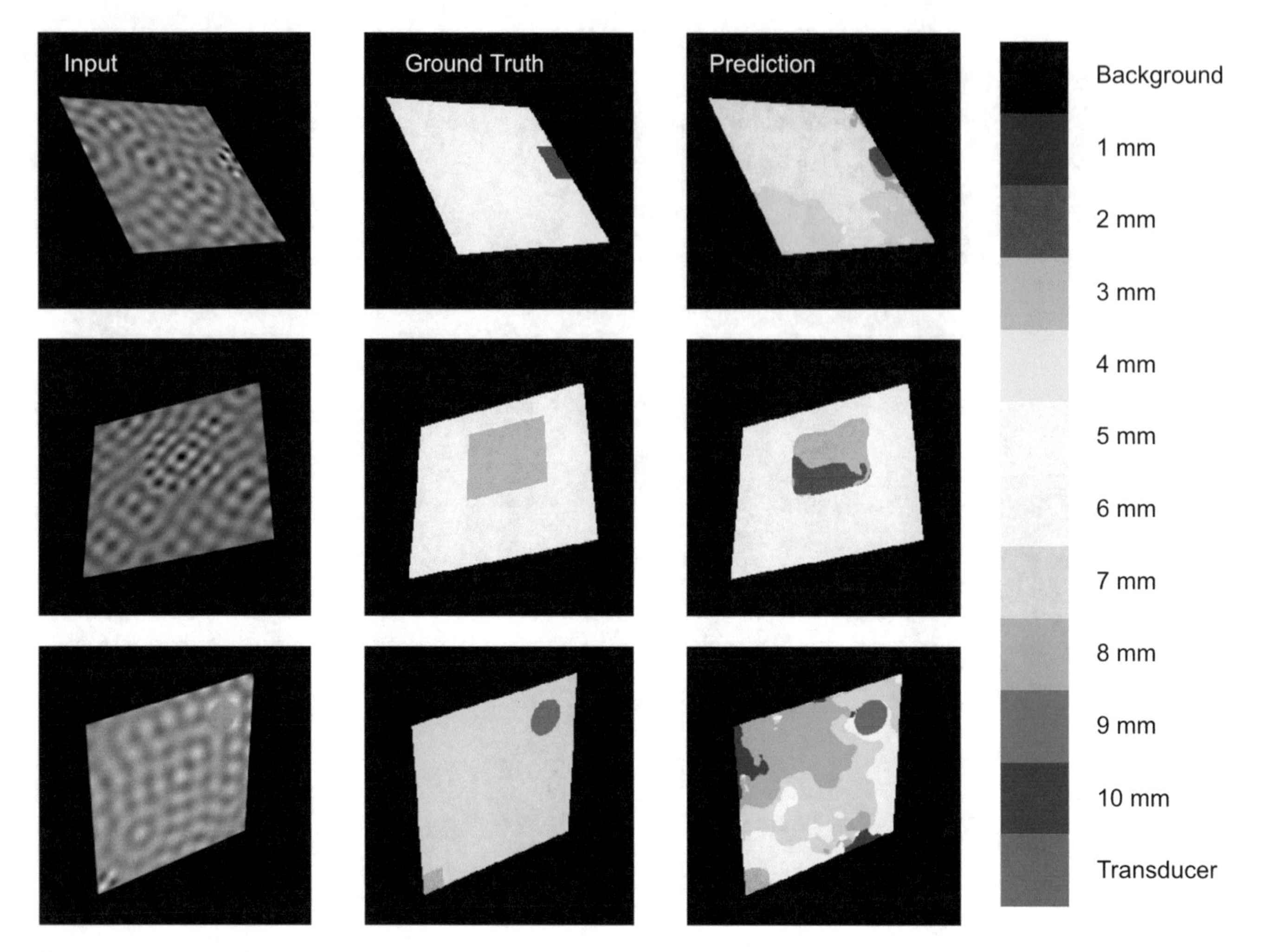

Fig. 17.5 Example set of CNN model predictions on test data with nominal thicknesses less than 10 mm. (Left) real component wavefield input (middle) ground truth label (right) CNN model prediction

was spatially sampled at a rate smaller than 1 mm and then interpolated onto a 1 mm^2 grid to match the resolution of the CNN input layer. The inspection surface was isolated from the full point cloud collected by the LiDAR and then aligned to the LDV response data and interpolated onto a 1 mm^2 grid. The CNN input channels collected experimentally were the real and imaginary wavefield response images and the radial distance of the focal point (LDV/LiDAR) to the inspection surface. Figure 17.9 displays the response data that has been interpolated onto a 1 mm^2 grid. Figure 17.10 shows the thickness map for the plate using a Fourier-based algorithm. The thickness map from this processing method contained artifacts at boundaries, and regional differences in thickness are not clearly outlined.

Figure 17.11 displays the model prediction for the experimental tilted plate data. The results reflect positively on the ability of the CNN to process information that is completely different from training/validation data. The leftmost image shows the cropped region of interest that simulates the zero edge boundary condition. The zero edge condition is valuable since it could represent a real-world situation where the structure is not a small plate with finite boundaries.

17.3.3 Temporally Shifted Snapshots

An additional study was conducted with experimental data that utilized temporally shifted snapshots at different instances within the steady-state sinusoidal excitation period. The test consisted of 37 wavefield images, with each one shifted by $\Delta t = 5/(180{\cdot}80\text{ kHz})$ increments starting at zero and ending at one full period of the excitation cycle. The plate used to test this method had a single curved defect of varying thickness which can be seen in the model predictions of Figs. 17.12

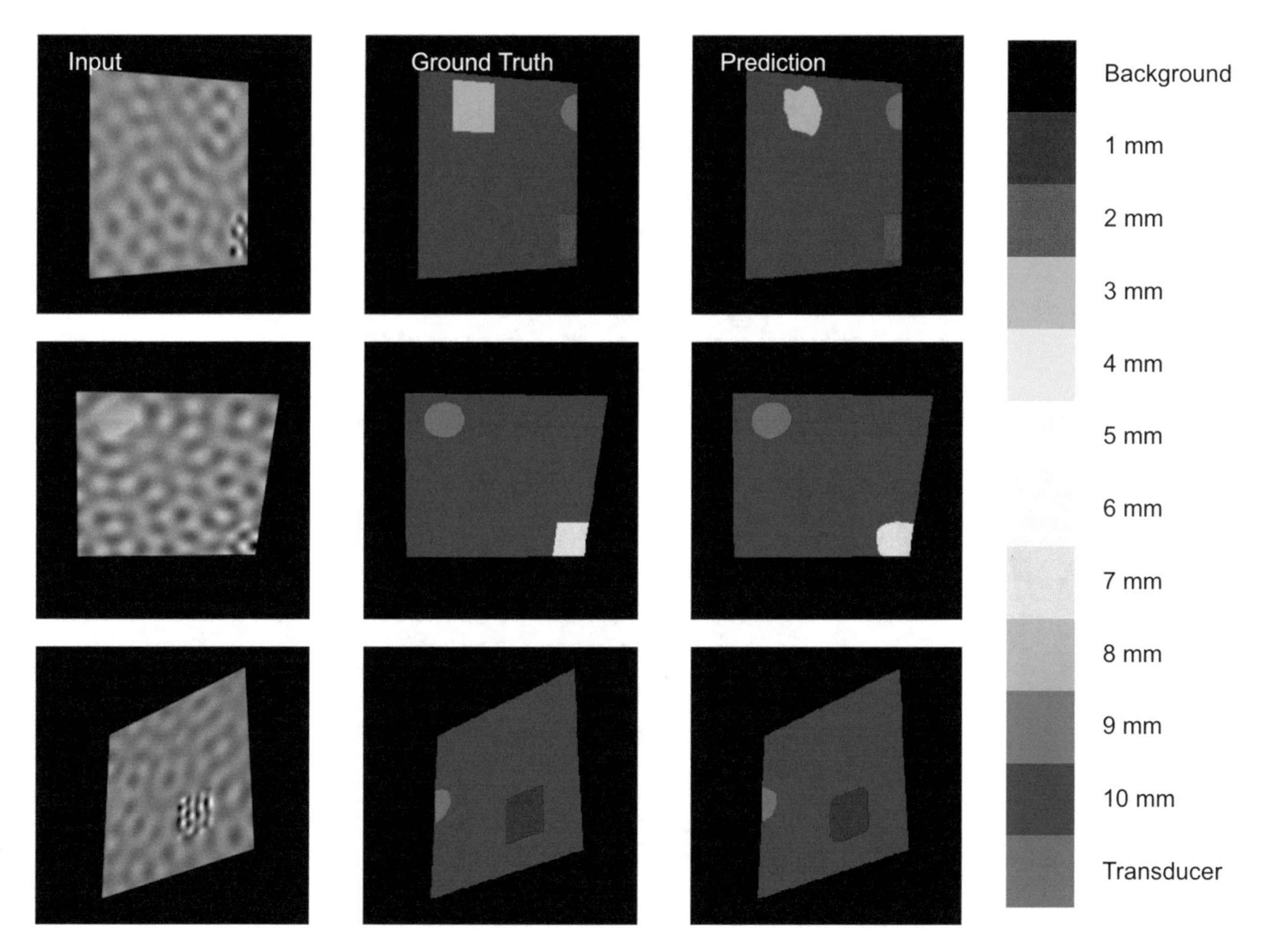

Fig. 17.6 Example set of CNN model predictions on test data with nominal 10 mm thicknesses

and 17.13. Predicting thickness for different temporally shifted snapshots of the surface response shows that there can be prediction variation depending on where the snapshot was taken within the excitation cycle. Each of the 37 snapshots were independently fed to the CNN model, and a composite prediction, Fig. 17.13, was computed from the average of all the prediction images. This methodology was observed to refine the predicted defect region as well as improve the accuracy of the thickness prediction throughout the plate.

17.4 Conclusion

This work has demonstrated how complex surface response data resulting from steady-state ultrasonic excitation and geometry mapping of the surface can be used to produce thickness maps for plate-like structures using a CNN. The inclusion of a geometry channel for the CNN that allows for application to surfaces that are not perpendicular to the LDV measuring the surface response shows the potential for this method to be applied to a wide range of structures and increases its usefulness in practical inspection situations. For tilted plates, the trained CNN model was able to successfully identify nominal thickness of the plate and locations of most defects without producing a significant number of boundary effects, such as those which are typically seen in other thickness estimation algorithms for these measurements. One key issue that should be addressed in future work is training the CNN with a more class-balanced training set to improve the performance of the CNN of the whole range of thicknesses. While choices like the use of a focal loss function were designed to address this issue, class balance provides a more robust solution approach. Changes to the model architecture may also be explored, such as the use

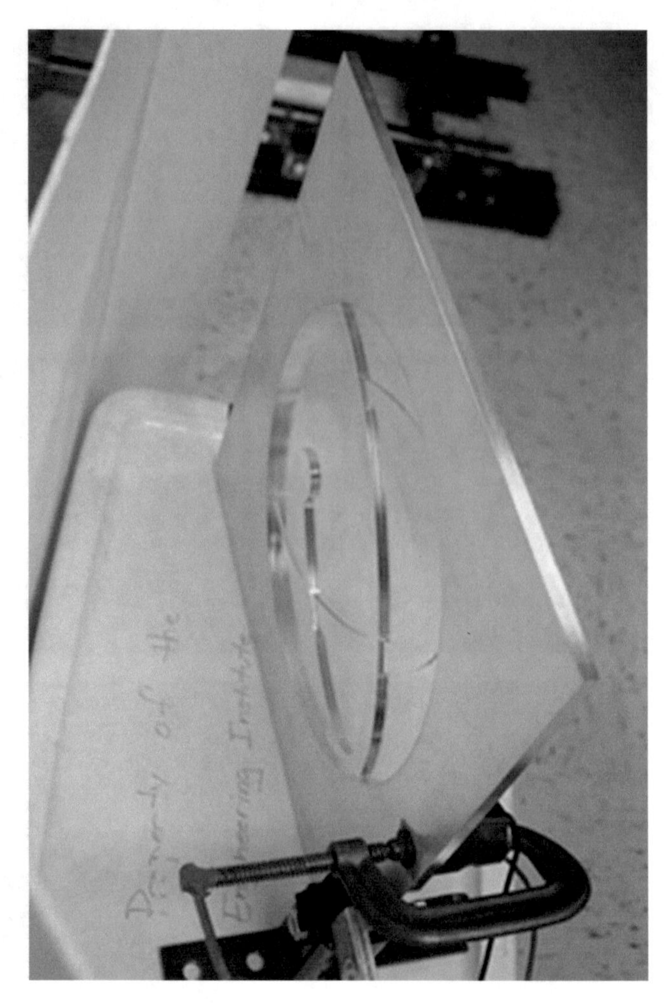

Fig. 17.7 Back view of experimental setup using a tilted plate

of a different pretrained encoder, like Resnet-50 or Resnet-101, or a different activation function, like LeakyRelu, which is known to prevent vanishing gradients in large models.

Acknowledgments The authors would like to thank Dr. Eric Flynn and Dr. Ian Cummings for their helpful discussions in regard to this research and Joshua Eckels, Isabel Fernandez, and Kelly Ho for providing the database of wavefield simulations. This research was funded by Los Alamos National Laboratory (LANL) through the Engineering Institute's Los Alamos Dynamics Summer School. The Engineering Institute is a research and education collaboration between LANL and the University of California San Diego's Jacobs School of Engineering. This collaboration seeks to promote multidisciplinary engineering research that develops and integrates advanced predictive modeling, novel sensing systems, and new developments in information technology to address LANL mission-relevant problems.

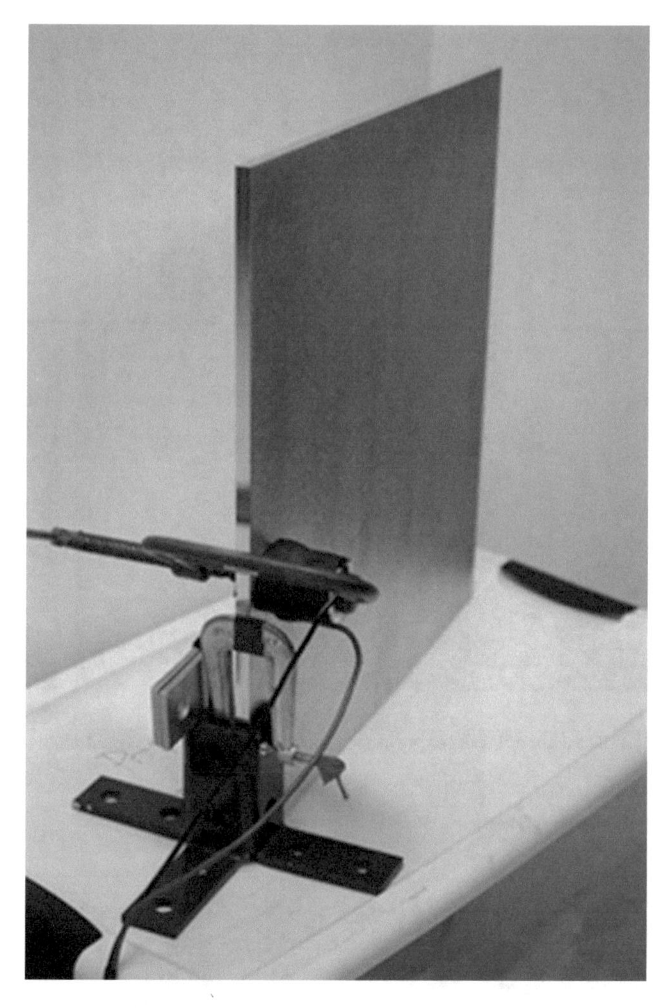

Fig. 17.8 Side view of experimental setup using a tiled plate. Surface response measurements were taken on flat surface shown

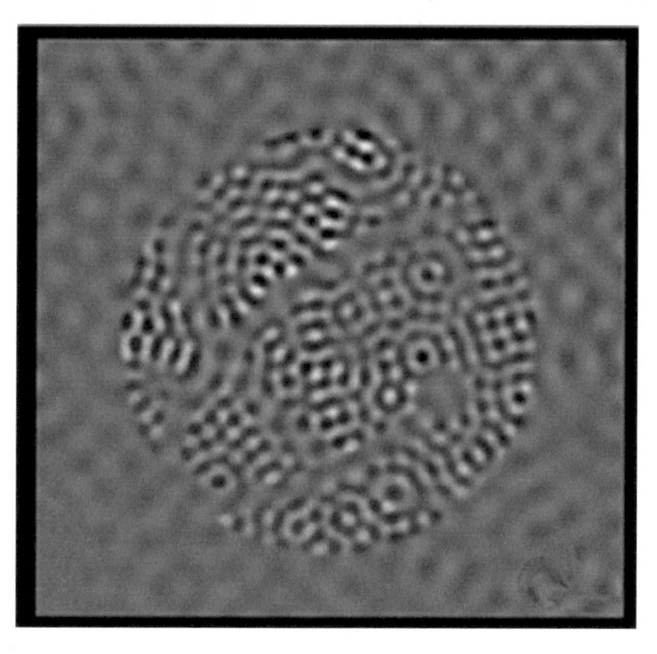

Fig. 17.9 Real component of response data for the experimental setup using the tilted plate

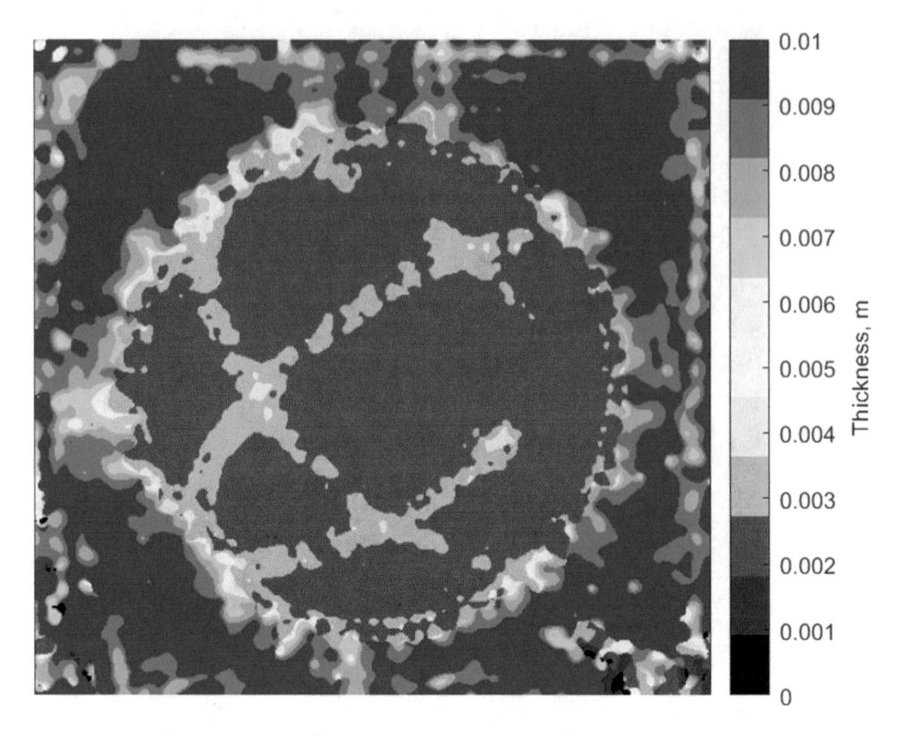

Fig. 17.10 Fourier-based ASSESS thickness estimation for experimental measurements on the tilted plate

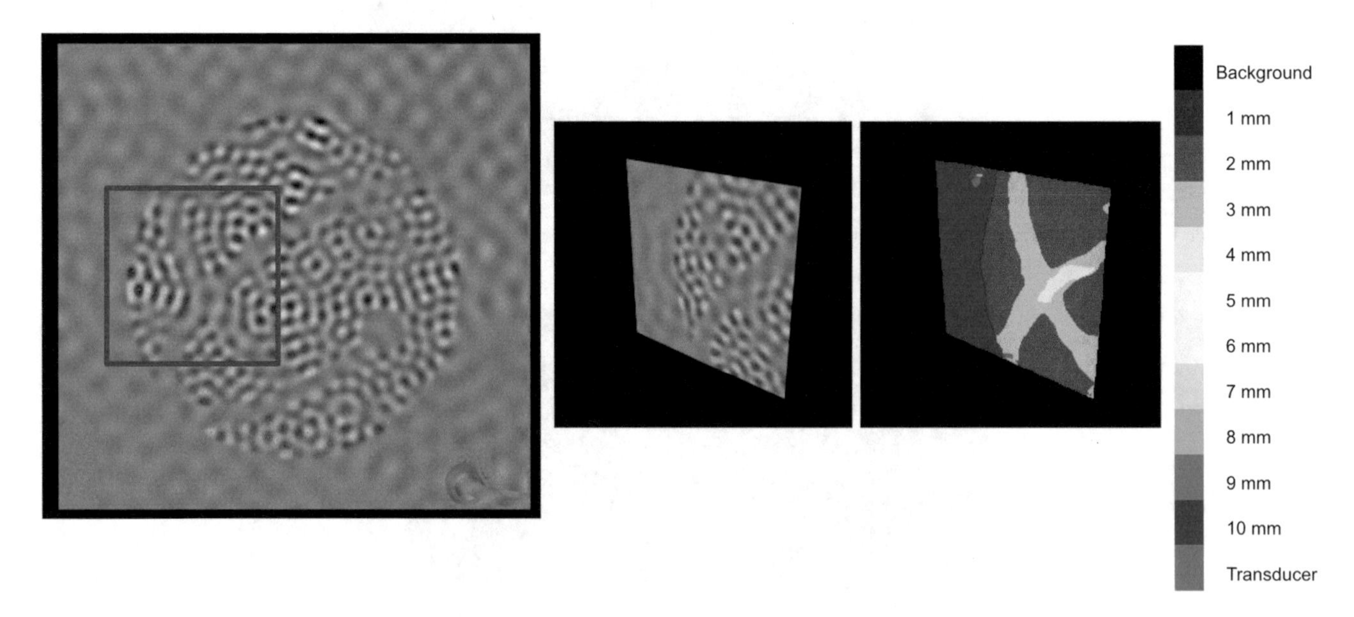

Fig. 17.11 (Left) red box indicates the region of interest for which titled plate data would be processed with CNN model (right) titled plate response and CNN thickness prediction

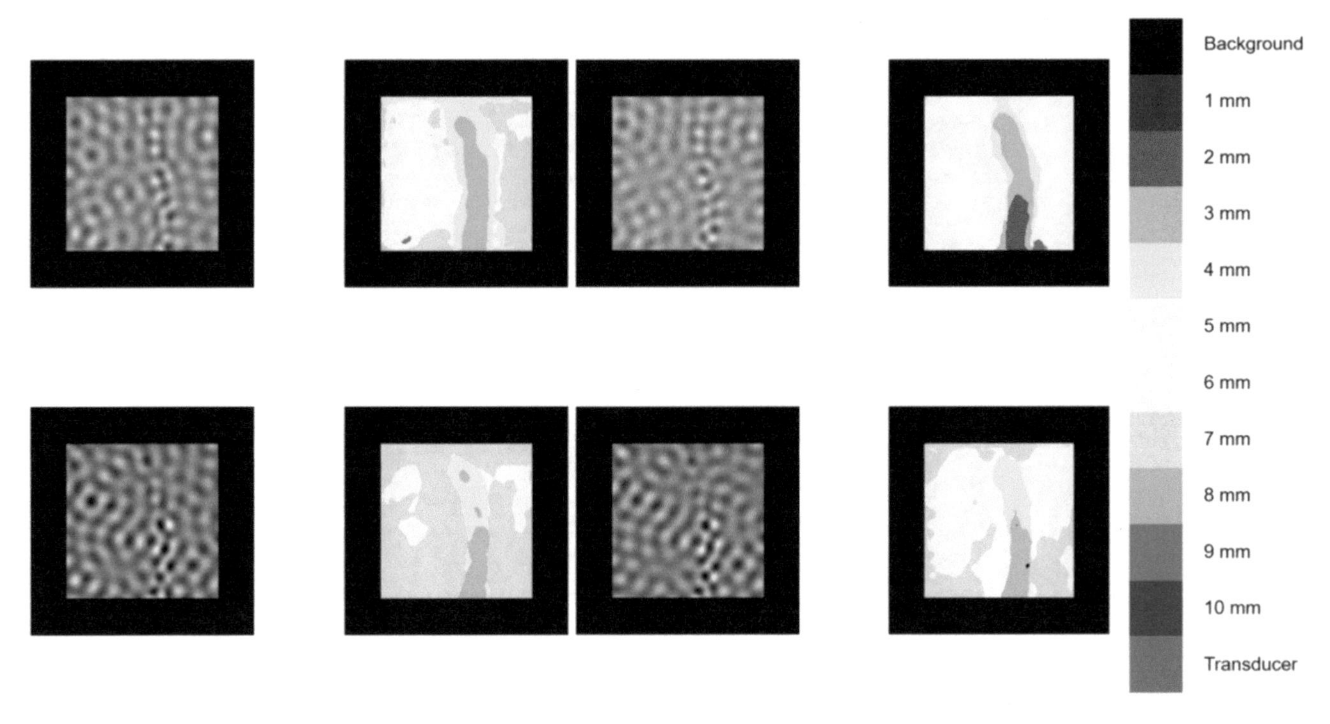

Fig. 17.12 Snapshot at phase shifts of 0°, 90°, 180°, and 270°(clockwise) with CNN predictions

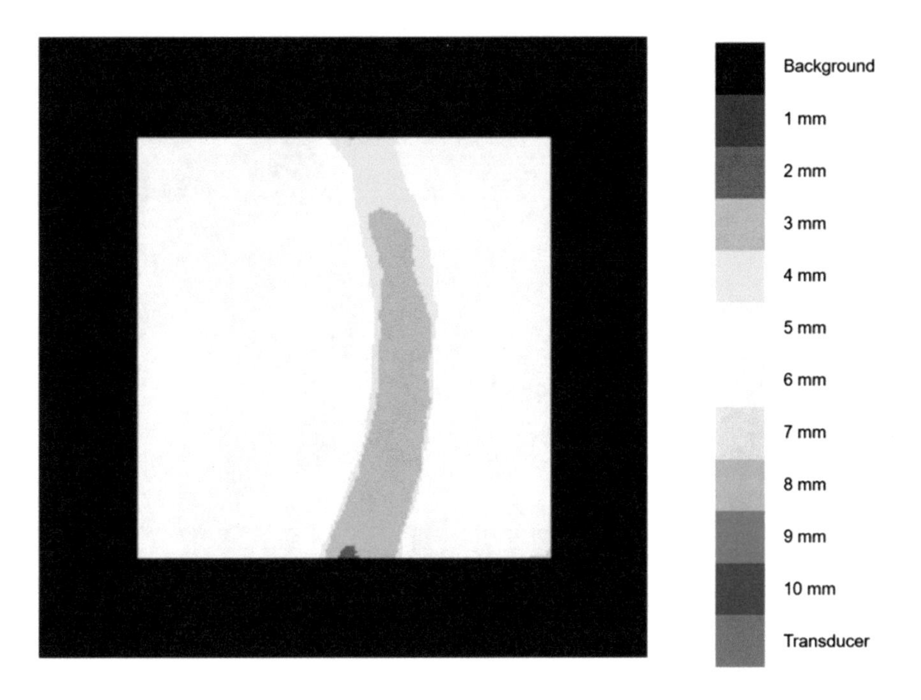

Fig. 17.13 Composite of all CNN predictions using every phase

Printed in the United States
by Baker & Taylor Publisher Services